Second Edition

AERIAL MAPPING

Methods and Applications

Second Edition

AERIAL MAPPING

Methods and Applications

Edgar Falkner
Dennis Morgan

CRC Press is an imprint of the
Taylor & Francis Group, an **informa** business

CRC Press
Taylor & Francis Group
6000 Broken Sound Parkway NW, Suite 300
Boca Raton, FL 33487-2742

First issued in paperback 2020

ISBN 13: 978-0-367-57872-5 (pbk)
ISBN 13: 978-1-56670-557-8 (hbk)

Visit the Taylor & Francis Web site at
http://www.taylorandfrancis.com

and the CRC Press Web site at
http://www.crcpress.com

Library of Congress Cataloging-in-Publication Data

Falkner, Edgar.
Aerial mapping methods and applications / Edgar Falkner and Dennis Morgan.--2nd ed.
p. cm.
Includes bibliographical references and index.
ISBN 1-56670-557-6
1. Photographic surveying. I. Morgan, Dennis (Dennis D.), 1951- II. Title.

TA593. F34 2001
526.9′82--dc21 2001038423
CIP

Library of Congress Card Number 2001038423

Preface

This book provides up-to-date information to help a variety of users; in particular, professionals and managers. The topics of photogrammetry, remote sensing, geographic information systems (GIS), global positioning systems (GPS), surveying, and other mapping sciences allow readers to develop a greater understanding, and help them harness the capabilities of technology.

Building upon principles described in the first edition of *Aerial Mapping: Methods and Applications*, the second edition captures current methods and describes their workings in language that is easy to understand. The growth in the mapping sciences over the last decade is breathtaking, and it is incumbent upon us to make use of the myriad advances. These technologies provide for more accurate and precise applications, and can often be conducted at lower relative cost than solutions provided by earlier technological approaches. With the details supplied here and your experience and skills, good things will happen.

John G. Lyon, Ph.D.
Henderson, NV

Foreword

Like its predecessor (*Aerial Mapping: Methods and Applications* which was written by Edgar Falkner and published by Lewis Publishers in 1995), this text was conceived to aid professional middle managers who may need to understand the rudiments of aerial photography, remote sensing, and photogrammetric mapping to get their job done. This may include, but not necessarily be limited to, the fields of agronomy, engineering, hydrology, surveying, geography, architecture, geographic information systems, soil science, forestry, wetlands, game biology, geology, natural resources, environmental science, public and private utilities, facilities management, or others.

The content of this text is deliberately semi-technical. It, in conjunction with the previous reference book, is intended as an introduction to practical mapping production. Parts of the original book have been included, but other sections have been upgraded in keeping with the dynamics of the technology.

Managers should supplement this knowledge with their own high level expertise. It should be realized that, as an elementary guide, the processes discussed are presented in a limited scientific context.

Until fairly recently photogrammetry was juxtaposed with such disciplines as geographic information systems, remote sensing, and image analysis. Now these fields all tend to merge. This is due, in great part, to advances in hardware and software for electronic data manipulation. In the current technological scenario many disciplines can share information. The reader must be aware that, parallel with the rapid advancement in electronic data processors, all of the technologies discussed herein are dynamic. What is vogue today could very well be obsolete tomorrow.

The original reference book essentially concentrated on photogrammetry, but this edition introduces the reader to other techniques that are available to aid in accomplishing a mission. Once exposed to basic procedures, the user can apply this knowledge to conditions that are germane to a specific project locale.

Edgar Falkner
Dennis Morgan
St. Louis, Missouri

The Authors

Dennis Morgan (right) has been a Civil Engineer and Certified Photogrammetrist with the U.S. Army Corps of Engineers since 1974. He earned a B.S. in Engineering Technology from Northern Arizona University in 1973. His professional background includes design, management, and monitoring of production of photogrammetric mapping data sets for a wide variety of engineering and Geographic Information Systems applications. His work has included aerial photography and planimetric, topographic, and orthophoto mapping for many civil and military engineering and environmental projects. Morgan is a member of the American Society of Photogrammetry and Remote Sensing (ASPRS). Published articles in various trade and professional journals include *POB,* Oct.-Nov. 1993, Vol. 19, No. 1, *The Military Engineer,* June-July Vol. 89, No. 585, and *GPS World,* Feb. 1996. He was a contributing author to *GIS DATA Conversion Strategies — Techniques — Management,* edited by Pat Hohl, Onward Press, 1998.

Edgar Falkner left high school during his senior year to serve a tour of duty with the Marine Corps during World War II. Upon discharge, he earned a B.S. in Forestry from Michigan College of Mining and Technology in 1953. For 7 years he held federal forest management (California) and state wildland inventory (Alaska) positions, and over a period of 34 years he worked in private and public aerial mapping situations, including 10 years as vice president and partner in a private sector firm and 5 years as a technical mapping consultant with the U.S. Corps of Engineers. He has frequently functioned as a lecturer for technical workshops, seminars, and adult education courses. He is the author of *Aerial Mapping: Methods and Applications*, published by Lewis Press in 1995. Falkner was a contributor to *GIS DATA Conversion Strategies — Techniques — Management*, edited by Pat Hohl, Onward Press, 1998, and is the author of several articles published in technical journals.

Acknowledgment

Within this book the reader will observe courtesy notes accompanying a number of illustrations, giving credit for contributions by several mapping and remote sensing organizations. In the past Ed and Dennis have worked on projects with representatives of these establishments as well as many of their contemporaries. During the course of these dealings the authors have found a common vein of helpfulness and cooperation at all times, and much of the information in this book has been gleaned from these people's free sharing of technical knowledge and professional experience. They deserve our thanks.

Contents

Chapter 10 Geographic Referencing

Chapter 11 Aerotriangulation

CHAPTER 1

Introduction

1.1 INTRODUCTION

From the Greek *photo* (light writing) *gram* (graphic) *metry* (measure) comes the root of the science of photogrammetry. The American Society for Photogrammetry and Remote Sensing (ASPRS) defines this methodology as "the art, science, and technology of obtaining reliable information about physical objects and the environment, through processes of recording, measuring, and interpreting images and patterns of electromagnetic radiant energy and other phenomena."

Photogrammetrists, individuals skilled in the application of photomapping procedures, produce maps directly from photographic images by identifying, symbolizing, and compiling elevational, cultural, and natural features that are visible on the imagery. Contours are created in their true terrain character, while planimetric features are located in their true horizontal positions. Image analysis is the art/science of interpreting specific criteria from a remotely sensed image.

ASPRS states that, "Remote sensing techniques are used to gather and process information about an object without direct physical contact." Remote sensors capture information from a source object that is significantly distant from the data collector.

In recent decades the advancement of image analysis, photogrammetry, and Geographic Information Systems has been due, in great measure, to progress in electronic data processing and remote sensing. Many aspects of these disciplines that were executed manually or mechanically a decade or two ago are now accomplished by analytical methods.

1.1.1 History of Photomapping

The birth of aerial photography was preceded by a lengthy development period.* Most maps that were produced in the past served a singular purpose, that of providing a visual planning tool to be utilized in a one-time design effort. During the 1980s

* For an in-depth discussion of the history of photomapping refer to Chapter 1 in *Aerial Mapping: Methods and Application,* Lewis Publishers, Boca Raton, FL, 1995.

and into the 1990s the progression of computers, stereoplotters, and map production systems moved at a hectic pace.

Currently, the field of photogrammetry is in the digital era, and today's photogrammetrist is more scientifically oriented than his yesteryear counterparts. Maps and data tabulations have become spatial data products in this age of increasing technological sophistication.

1.1.2 Photogrammetrists and Image Analysts

Both photogrammetrists and image analysts make extensive use of aerial photographs to prepare maps.

1.1.2.1 Photogrammetrists

Photogrammetrists must develop the capability to interpret photo image features, relate them to ground equivalents, and orient them to a prescribed spatial datum. They must learn to identify cultural features on a photo image so that feature details can be correctly symbolized into recognizable map features. They must also have some recognition of landforms so that the elevation data correctly depict the terrain. Most of these individuals gain proficiency by translating the photo image into what they have previously observed on the ground during their daily routine pursuits.

1.1.2.2 Image Analysts

Image analysts do much the same as photogrammetrists, but from a different point of view. They must apply the knowledge gained from a technically oriented background in specific disciplines. A few examples may clarify this concept:

- A soils specialist may be looking for erodible soils on a proposed highway route.
- A forester may be estimating the volume on a timber tract.
- An entomologist may be attempting to discern the prevalence of corn blight in a specific township.
- A hydrologist may be comparing the degree of suspended matter in several adjacent lakes.
- A wetland specialist may be monitoring the decline of wetland areas within a state.
- A wildlife biologist may be inventorying migratory geese on a wildfowl refuge.
- A glaciologist may be charting the movement of a glacier.

1.1.3 Utilization of Aerial Photos

The photogrammetrist uses photographs to directly create an end product such as a planimetric and/or a topographic map, while the analyst uses the photographic image only as one implement in a variable toolbox to arrive at a product.

1.1.3.1 End Products

A photogrammetrist generates an end product, graphic or digital, directly from the photographs by identifying, symbolizing, and compiling cultural and terrain features that are visible on the imagery. Usually, a limited effort is directed toward field verification of the product. In other cases, an image analyst proceeds through various phases, combining image analysis with ground truth sampling to produce an end product.

1.1.3.2 Effort

Major portions of a photogrammetrist's efforts are by direct use of the photos. Conversely, the analyst's use of images may not be a major effort in the project scheme.

1.1.4 Photogrammetry

Several types of photogrammetry exist: aerial, terrestrial, and close range. Each serves the needs of a distinct category of users. Throughout the mapping community, terrestrial and close range photogrammetry have limited use. Aerial photogrammetry uses near-vertical photographic images that are exposed from a moving platform at a distant point in the sky. This procedure is employed to develop planimetric detail and/or topographic configuration. Aerial photogrammetry is also employed for numerous aerial photo analysis purposes.

1.1.4.1 Digital Mapping

Digital planimetric and/or topographic mapping projects require at least several basic operations that include acquisition of:

- Aerial photography
- Field control surveys
- Digital data collection and attribution

1.1.4.2 Supplemental Functions

Aerial mapping projects often necessitate supplemental functions such as:

- Aerotriangulation
- Photographic reproduction products
- Orthophoto mapping
- Accessory field surveys such as outboundaries, cross-sections, drill hole locations, or utilities information

1.1.4.3 Commercial Mapping

The web site http://www.aeromap.com/ provides insight on the equipment, services, applications, and representative projects of a commercial aerial mapping firm.

1.1.5 Mapper vs. User

Most map users contract with mapping firms to accomplish the actual production procedures. Mapping requires the use of expensive equipment, and map production is labor intensive. Hence, mapping can be a costly venture. Similarly, users are often confined to a strict budget that may or may not be sufficient to cover the cost of suitable mapping. There are times when the user might wish, but cannot expect, to obtain a map fulfilling all of the project needs for the amount of funding available. These situations could provoke a somewhat adversarial situation. The mapper must exercise professional integrity to produce a quality product in keeping with the needs of the user, but the user must be willing to pay a realistic price for that credible product.

CHAPTER 2

Electromagnetic Energy

2.1 RADIANT ENERGY

To understand the concept of aerial photography and other remote sensors, the mapper, map user, or image analyst must have at least a nodding acquaintance with radiant energy (flux).

2.1.1 Radiant Waves

All forms of radiant energy, which are components of the electromagnetic spectrum, travel in waves similar to those illustrated in Figure 2.1. Electromagnetic energy involves two functions: frequency and wavelength. Frequency is the rate at which the oscillations pass a given point. Wavelength is the distance between any point on one wave and the analogous point on the next wave. Velocity of electromagnetic energy, a constant in a vacuum, is the product of these factors, which are both variables. In this chapter, only the wavelength is mentioned in the discrimination of various spectral regions (ultraviolet, visible, infrared, microwaves).

Radiant waves are deflected by colliding with any foreign particle of matter which is larger than that wavelength. The shorter the wavelength, the more it is scattered by dirt particles, water droplets, vapor, and gas in the air.

2.1.2 Distribution of Energy

The sun emits solar energy that permeates the earth, and objects on the surface absorb, transmit, and/or reflect varying amounts of solar energy (Figure 2.2). Aerial films are sensitive to visible light waves that reflect from these objects. Some specialty films react to near infrared radiation.

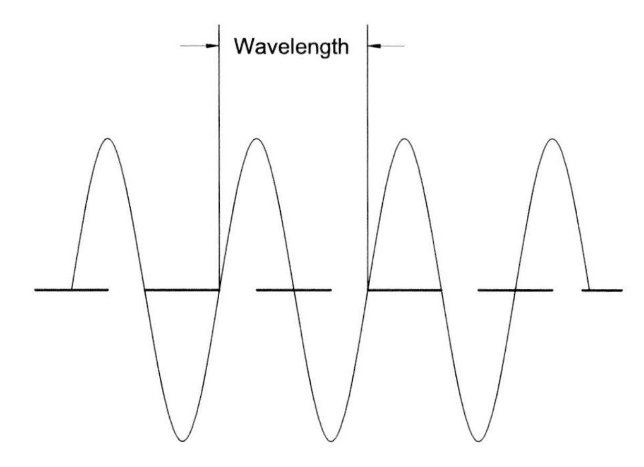

Figure 2.1 Radiant energy waves.

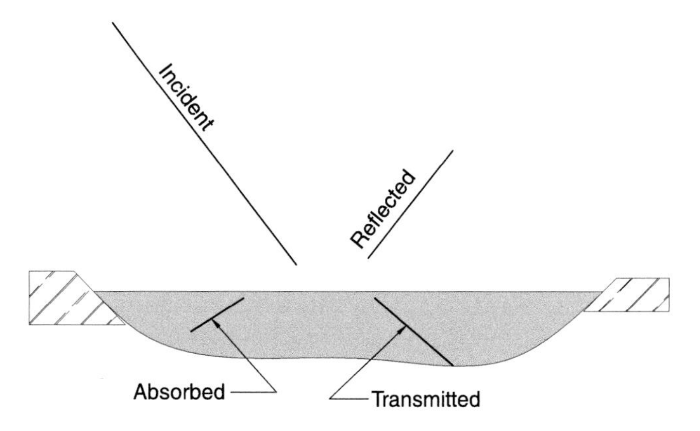

Figure 2.2 Distribution of solar energy.

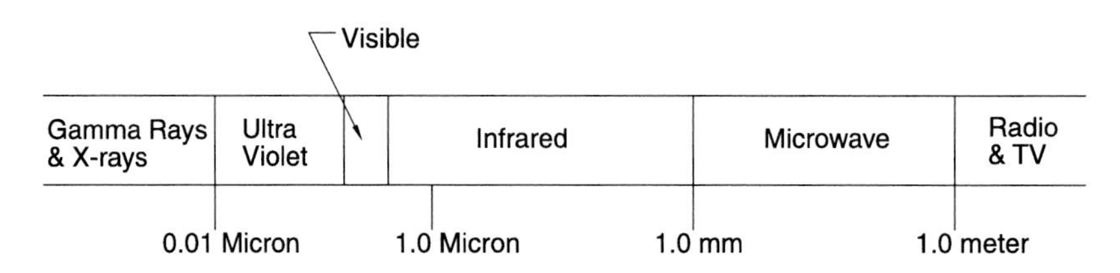

Figure 2.3 Components of the electromagnetic spectrum.

2.2 ELECTROMAGNETIC SPECTRUM*

Figure 2.3 represents the electromagnetic spectrum, indicating component energies detected by remote sensors in their relative wavelength regions. Elements of

* Internet keyword "electromagnetic spectrum" leads to a number of illustrated perspective views of radiant energy.

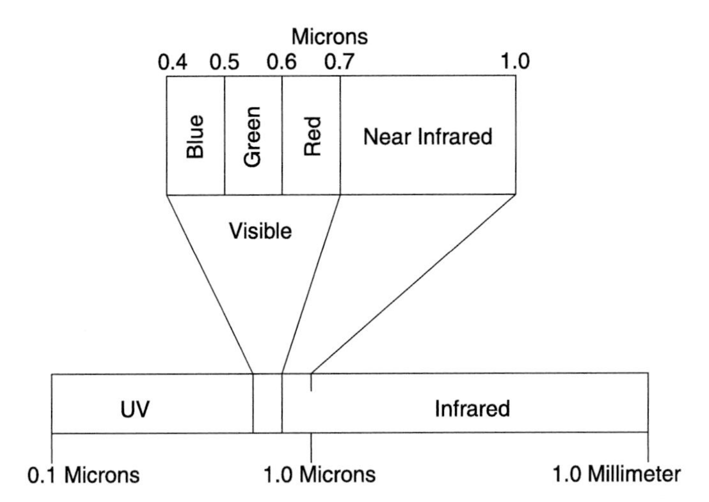

Figure 2.4 Basic components of visible light and their relationship with shorter and longer wavelengths.

the electromagnetic spectrum that are currently utilized in photogrammetry and remote sensing are the ultraviolet to a limited extent along with visible, infrared, and microwave to a considerable extent.

At the lower end of the portion of the spectrum utilized by remote sensors, wavelengths are measured in fractions of micrometers (μm). At the upper range measurements are in millimeters (mm) to centimeters (cm) and are termed microwaves.

2.2.1 Visible Light

Human eyes see only that portion of the electromagnetic spectrum denoted as visible energy. This visible energy makes up a very small portion of the radiance that is scattered around the sky at any point in time.

Humans see the visible portion of the electromagnetic spectrum as various colors that span a range encompassing 0.4- to 0.7-μm wavelengths. Various colors of the rainbow are blends of the additive primary physical colors of red, green, and blue.* Visible light is sensed by a camera or multispectral sensor. Figure 2.4 visualizes the basic components of visible light and their relationship with shorter and longer wavelengths.

Blue wavelengths are the shortest of the visible light portion of the spectrum, and they ricochet off the most minute particles of gas and vapor, causing them to disperse all over the sky. Green and red are longer and are deflected by minuscule particles of dust and water droplets.

This prolific scattering of the shorter waves dominates the sense of vision and compels humans to see a blue sky. However, as the size of the particulate matter increases — caused by smoke, moisture, or dust storms — the longer waves are forced to rebound. Thus, more of the greens and reds fill the sky.

* Equal parts of blue, green, and red appear as white light. The absence of all three results in black.

2.2.2 Infrared

Infrared may be a confusing term, but it simply refers to heat radiation. Two types of radiation will be detectable by specific sensors: reflected and emitted. Both may be used to good advantage by the image analyst.

2.2.2.1 Reflected Infrared

Near infrared, or reflected infrared, refers to the shorter infrared wavelengths and indicates the relative amounts of solar radiation which reflect off the molecular composition of the surface of an object. Near infrared radiation fits into the electromagnetic scheme in the 0.7- to 1.1-μm wavelength range. Reflected infrared does not indicate the actual temperature of the mass.

Reflected infrared can be detected by a camera. Examples of this application would be:

- Healthy vegetation (whether leaves on trees or bushes, blades of grass, or foliage of cultivated crops) produces sugar through the process of photosynthesis. When this chemical function decreases or stops, the leaf surface takes on a modified molecular structure. The amount of infrared reflection differs at these various stages and is seen as different hues. This effect is especially striking with color infrared imagery, where healthy vegetation appears red and various stages of lesser vigor result in more subdued pinks.
- Clean water absorbs near infrared waves, so this image tends to be very dark on infrared images. As the amount of suspended particles increases, the infrared waves collide with this foreign material and are reflected, resulting in a lighter image tone.

2.2.2.2 Emitted Heat

Thermal infrared, or emitted heat, wavelengths are longer, contained within the 1.0- to 13-μm band, and denote actual temperature radiation emitting from an object. These fall into two categories: middle infrared and far infrared. A thermal scanner rather than a camera must sense emitted heat images. For example, if an infrared scanner were aimed at a house in winter it would sense heat leakage from poorly insulated areas on the surface of the building by exhibiting a greater radiometric value on the image than that of the dark return of the colder background. Appropriate computer instrumentation breaks this information into variable light intensity pulses that are used to create the pictorial image.

CHAPTER 3

Aerial Films

3.1 AERIAL FILMS

Aerial film is similar in construction to the film popularly used in handheld 35-mm cameras. It comes in rolls that are 10 in. wide and range in length from 200 to 500 ft. Figure 3.1 depicts the basic structure of aerial film.*

3.1.1 Types of Film

Although there are a number of aerial films in use, many serve unique situations. Two commonly utilized films employed in planimetric and/or topographic digital mapping are panchromatic and natural color. These two films plus infrared and false color form the basic media used in image analysis procedures.

*3.1.1.1 Panchromatic***

Panchromatic, more often termed black and white, is the most commonly encountered film employed for photogrammetry. The sensitive layer consists of silver salt (bromide, chloride, and halide) crystals suspended in a pure gelatin coating which sits atop a plastic base sheet. Visible light waves react with the silver particles in the emulsion, causing a chemical reaction that creates a gray-scale image. The emulsion is sensitive to the visible (0.4- to 0.7-μm) portion of the electromagnetic spectrum that is detected by the human eye.

*3.1.1.2 Color****

Natural color film is also called true color or color. The multilayer emulsion is sensitive to the portion of the electromagnetic spectrum that is visible to the human

* For a broader discussion of film components refer to Chapter 3 in *Aerial Mapping: Methods and Applications,* Lewis Publishers, Boca Raton, FL, 1995.
** Refer to http://www.kodak.com/US/en/government/aerial/products/film/blackWhite.shmtl on the Internet.
*** Refer to http://www.kodak.com/US/en/government/aerial/products/film/color.shmtl on the Internet.

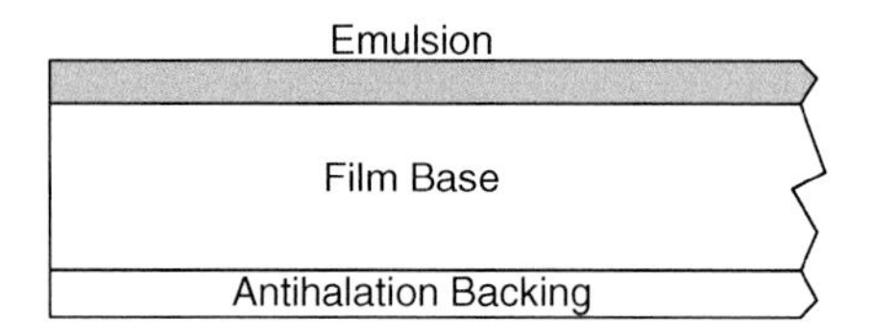

Figure 3.1 Basic components of panchromatic film.

eye. There are three layers of gelatin containing sensitized dyes, one each for blue (0.4–0.5 μm), green (0.5–0.6 μm), and red (0.6–0.7 μm) light. Green and red layers are also sensitive to blue wavelengths. Visible light waves first pass through and react with the blue layer and then pass through a filter layer which halts further passage of the blue rays. Green and red waves pass through this barrier and sensitize their respective dyes, causing a chemical reaction and thus completing the exposure and creating a true color image.

*3.1.1.3 Infrared**

Current aerial infrared film is offered as two types: black and white infrared and color infrared.

Black and White Infrared

Infrared film is also known as black and white infrared. The emulsion is sensitive to green (0.54–0.6 μm), red (0.6–0.7 μm), and part of the near infrared (0.7–1.0 μm) portions of the spectrum and renders a gray-scale image. Positive images appear quite like panchromatic film, except that water and vigorous vegetation tend to register as darker gray to black. The film structure resembles panchromatic with the exception that the emulsion sensitivity range is shifted upward, eliminating blue wavelengths and including a portion of the near infrared. In the past this film was used extensively in vegetation and water studies, but its popularity seems to be declining in favor of color infrared.

Color Infrared

Color Infrared film is commonly termed false color. The multilayer emulsion is sensitive to green (0.5–0.6 μm), red (0.6–0.7 μm), and part of the near infrared (0.7–1.0 μm) portions of the spectrum. A false color image contains red/pink hues in vegetative areas, with the color depending upon the degree to which the photosynthetic process is active. It also images water in light blue/green to dark blue/black hues, depending on the amount of particulates suspended in the water body. Clean water readily absorbs near infrared radiation. As the amount of foreign particulates

* Refer to http://www.kodak.com/US/en/government/aerial/products/film/infrared2443.shtml on the Internet.

increases, the near infrared rays reflect increasingly more of these particles. The film structure resembles natural color, except that the blue sensitive layer is eliminated and replaced by a layer that reacts to a portion of near infrared (0.7–1.0 μm).

3.2 SENSITOMETRY

Sensitometry is the science pertaining to the action of exposure and development on photographic emulsions.

At the instant of exposure, flux passes through the emulsion, causing minuscule silver salt crystals to be chemically converted into metallic silver. The number of transformed silver crystals is high in areas that are exposed to a great light intensity. Conversely, in areas of lesser exposure the amount of converted silver is restricted.

When a light source is passed through a developed negative, areas appear where much light passes through the film. There are also areas where small amounts of light pass through this medium. These gradations of metallic silver concentrations, which render portions of the film transparent to opaque, are manifestations of density. This montage of variegated densities creates a latent image. Relative amounts of density can be measured with a densitometer, which is an instrument that senses the proportion of a projected light beam passing through the film. Image contrast is the distinct discrimination of these various densities.

3.3 FILTERS*

Aerial photographs are usually exposed through a glass filter attached in front of the lens, so as to enhance the image in some fashion. Filters absorb unwanted portions of the spectrum to enhance image quality by reducing problems such as haze or darkening of the image at the edge of the exposure. There are a variety of filters that can be employed depending upon the type of film and the purpose of the imagery.

3.4 FILM PROCESSING**

Aerial film is developed in automatic processing machines, where the exposed film enters one end and the processed negative exits the other. Chemical temperatures and development timing sequences are critical, more with color as compared to panchromatic films. These thermal ranges and temporal periods during which the film is immersed in the various liquid baths should be regulated per specifications established by the manufacturer.

* For a broader scope of filters refer to Chapter 3 in *Aerial Mapping: Methods and Applications,* Lewis Publishers, Boca Raton, FL, 1995.

** For a broader scope of film processing refer to Chapter 3 in *Aerial Mapping: Methods and Applications,* Lewis Publishers, Boca Raton, FL, 1995.

3.5 RESOLUTION

Two terms, definition and resolution, are often used interchangeably when discussing images:

- Definition is the clarity of the image detail.
- Resolution is the size of the smallest unit of data that forms the image.

In the field of photogrammetry and remote sensing, resolution is thought of as resolving power. When referring to sensed images, resolution has several connotations: spectral, spatial, and radiometric.

3.5.1 Spectral Resolution

Spectral resolution is the wavelength band to which a sensor is sensitive. For example, natural color film is sensitive to blue, green, and red visible colors, a bandwidth spanning 0.4–0.7 μm. Wavelength bands should be selected so as to produce the best contrast separation between an object and its background.

3.5.2 Spatial Resolution

Spatial resolution is the smallest unit which is detected by a sensor. In a scene created from data captured by a resource satellite, the resolution may be a pixel (picture element) that is 15 m^2. By way of contrast, the resolving power on an aerial photograph may be 50 line pairs per millimeter.

3.5.3 Radiometric Resolution

Radiometric resolution is the sensitivity of a detector to measure radiant flux that is reflected or emitted from a ground object. For instance, full sunlight reflecting from the metal roof of a building will register as a brighter intensity than from a dark-shingled roof.

3.6 APPLICATION OF AERIAL FILMS

Dual camera systems can be used to expose two different types of film simultaneously. This might be the case where it is desirable to obtain natural color and false color imagery for a vegetation stress study.

Although some films are processed to a positive form, no opportunity exists yet to produce positive paper prints from these exposures. When processed to a positive form, natural and false color films offer advantages over paper contact prints. Film transparencies provide a first-generation product that enhances the definition as well as the image analysis capabilities. Transparencies can be studied with appropriate stereoscopic viewing devices over a light table while the film is on the roll. This procedure provides abundant image backlighting as well as ease of film handling.

Panchromatic and natural color enlargements are in great demand by the general public. These are used for such purposes as wall hangings, site promotion, product display, court trials, accident scene records, and informational presentations. Color infrared enlargements are often required by the image analyst or photogrammetrist for promotional or illustrative purposes.

3.6.1 Panchromatic

Panchromatic film can be used for mapping or image analysis and is developed to a negative form. Applications for this type of film require the use of contact prints and/or film diapositives. Diapositives are film plates that are utilized in stereocompilation.

3.6.2 Infrared

Infrared film is employed in image analysis and is developed to a negative form. Applications for this film require the use of contact prints and film positives.

3.6.3 Natural Color

Natural color film is utilized to some extent in planimetric and/or topographic mapping, but not as much as panchromatic film. If used for mapping, natural color is developed to a negative. Then contact prints and diapositives are produced in positive form. If the exposures are to be used in image analysis, some natural color films can be processed to either a positive transparency or a negative form. Color film, when processed to a negative, exhibits subtractive color primaries of cyan, yellow, and magenta. When processed to a positive, the additive color primaries of blue, green, and red are exhibited.

Figure 3.2 illustrates how the color image is processed to a negative and then produced as a positive.

3.6.4 Color Infrared

Since there is no blue-sensitive layer on color infrared film, each layer is "bumped up" to the next wavelength bandwidth. Refer to Figure 3.3 to understand this color transferal process. This shift in color coding of emulsions is the reason green foliage appears as a red image and exposed red clay is rendered as a green tone. Color infrared is usually developed to a positive transparency. The transparency can be used directly in stereocompilation or image analysis. This provides the benefits inherent in first-generation image definition, which may include sharper detail and ease of interpretation.

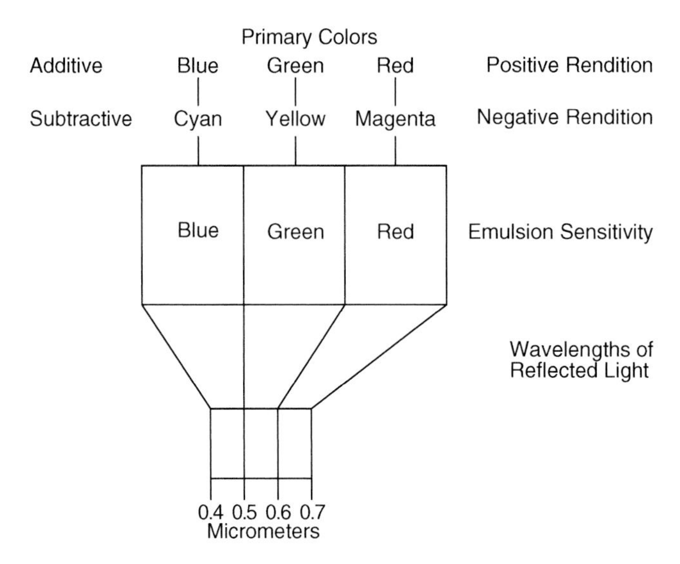

Figure 3.2 Spectral light compared with negative and positive color rendition on natural color films.

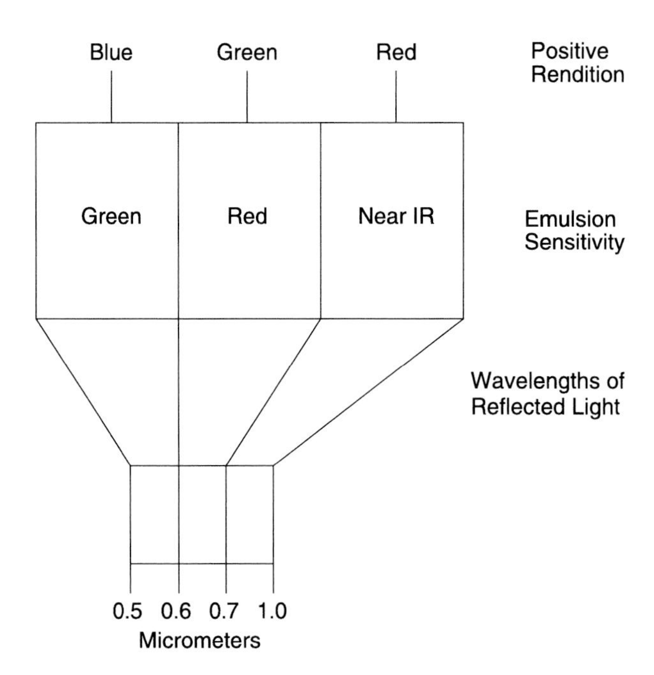

Figure 3.3 Spectral reflections compared with positive color primaries on false color film.

CHAPTER 4

Airborne Sensor Platforms

4.1 INTRODUCTION

Historically, photogrammetric applications have relied upon aerial photographs as a basic tool. Aerial photography, as the name suggests, requires the use of an airborne platform from which to expose the film. Although airplanes, helicopters, and lighter-than-air craft are employed as aerial photography vehicles, fixed-wing aircraft are the primary aerial photographic platform. Recent technological advances have changed aerial image collection drastically. Historically, aerial images have been collected with the aid of analog camera systems, but technology advances in digital camera systems have made great strides in recent years. Single image digital cameras and digital videography have a place today in photogrammetry. Near future technology advancements will bring these camera systems to the forefront. Global Positioning System (GPS) techniques, digital cameras, and image motion units may improve the accuracy of image collection and speed future mapping processes by collecting images and their location data in near real time. The basic platforms have remained unchanged, but the collection options are undergoing amazing changes.

4.2 FIXED-WING AIRCRAFT

A conventional fixed-wing aircraft, which allows for flexible schedules and cost-effective data collection, is the usual airborne sensor platform of choice. The type of aircraft to be used depends upon the requirements of the data collection.

4.2.1 Single-Engine Platform

Many projects require large-scale imagery over a relatively small area. The commute to and from the project site is often short, and the image collection mission can be completed in less than a single day's available flight time. Many of these types of projects can be accomplished with a single-engine aircraft when the required

Figure 4.1 Twin Otter aircraft, typical aerial photographic platform. (Courtesy of Surdex Corporation, Chesterfield, MO.)

altitude of the flight remains below about 18,000 ft above mean sea level. Image collection projects that require altitudes above this altitude and/or require longer commutes may mandate a more powerful multiengine airplane.

4.2.2 Multi-Engine Platform

Twin-engine aircraft, such as the Twin Otter seen in Figure 4.1, can also be successfully operated at lower altitudes. Multiple engine aircraft are normally faster, but are also more expensive to amortize, operate, and maintain. These aircraft also generally have the ability to carry more equipment, which may allow for more than one camera port in the aircraft and multiple, simultaneous data collection. The higher cost of these aircraft must be redeemed by project requirements, such as long commutes, higher altitude requirements, and large project areas requiring massive amounts of image collection, and/or multiple image type requirements.

4.3 FLIGHT CREWS

A variety of aircrew and camera system arrangements are possible. An aircrew can be comprised of one, two, or three persons, and the aircraft may have the capability to simultaneously collect data from more than a single image collection system. One example would be to collect both panchromatic and color infrared imagery of the same site at the same time, which would require two aerial cameras mounted in different ports in the same aircraft. Another example would be to collect digital multispectral data and panchromatic imagery of the same site at the same

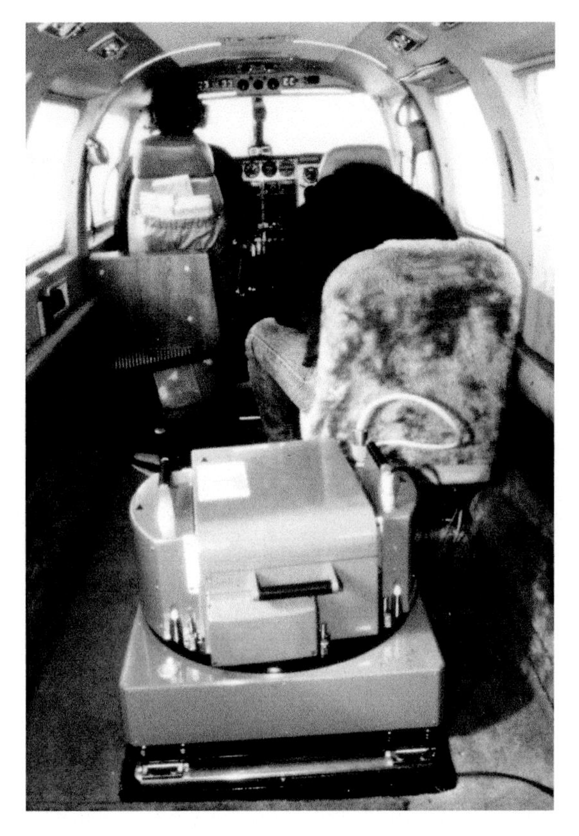

Figure 4.2 Typical aerial photographic installation. (Courtesy of Surdex Corporation, Chesterfield, MO.)

time. The purpose of this setup might be to collect and analyze vegetation health from the multispectral data and georeference it with planimetric and topographic mapping most economically obtained from panchromatic film.

Some setups today allow the pilot to safely navigate the aircraft as well as operate and monitor multiple image and sensor collection devices. Most aerial sensor platforms use a two-man crew consisting of a pilot and a photographer, such as the setup shown in Figure 4.2. The pilot flies and navigates the airplane, while the photographer ensures that the camera system is functioning properly.

Advances in airborne GPS navigation, digital cameras, and image motion units are changing the technical complexity and duties of the aerial photography crew. The pilot often navigates the aerial mission by following a predetermined path programmed on a video screen. Aerial images are collected with either an analog metric aerial camera or a digital camera system. The second crewmember is often an image specialist with specific knowledge of analog aerial cameras, film, and digital camera systems, including computer software and hardware for collecting and storing images and GPS surveying.

Figure 4.3 Electronic preflight planning station. (Courtesy of Surdex Corporation, Chesterfield, MO.)

4.4 NAVIGATION

Preplanning a photographic mission includes establishing the project area, the flight height, and the required number and position of the flight lines.

Aircraft navigation along photographic flight paths can be by visual recognition, where the pilot guides the flight of the aircraft by visually following identifiable ground features. Perhaps more commonly, electronic guidance methods are employed which are based upon GPS technology. GPS aircraft navigation systems for aerial image collection require electronic preplanning of the project. An electronic flight preplanning station is depicted in Figure 4.3. Once the approximate coordinates of the beginning and ending frames in each flight line are calculated and input into the flight planning software, the system allows the pilot to view and monitor the flight path on a computer screen during the mission. This type of system ensures complete image collection of a project in an efficient preplanned configuration.

4.5 HELICOPTER PLATFORMS

A limited amount of specialized mapping photography is acquired using a helicopter as a platform. Helicopters can be put to use with great advantage in situations where very large-scale photography or videography is required. Reduced speed and hovering capabilities of a helicopter allow the photographer to better compose the camera perspective than with an airplane. Large-scale mapping, oblique photography, and reconnaissance imagery can be accomplished for small area projects with a helicopter platform. The largest limitation of this platform type is the fuel capacity of the vehicle, since helicopter platforms generally have less fuel capacity than fixed-wing aircraft platforms, which translates to longer mission times and higher costs for larger projects.

4.6 AERIAL CAMERAS

Several precision aerial mapping camera systems are on the market; all are very expensive because of rigidly controlled construction and meticulous lens polishing. These cameras are finely adjusted precision photographic instruments. Aerial camera systems include both analog and digital cameras.

4.6.1 Camera Mount

For an aircraft to operate as a high-quality photographic platform, it is necessary to cut a hole in the bottom of the aircraft. A camera mount is attached to the floor and centered over the hole, and the camera then slips into this mount. A camera can rotate horizontally and be tilted several degrees in two directions, permitting compensation for the inconsistencies in the flight attitude.

4.6.2 Analog Camera Components

A cross-sectional representation of the pertinent components of a typical analog aerial camera is shown in Figure 4.4.

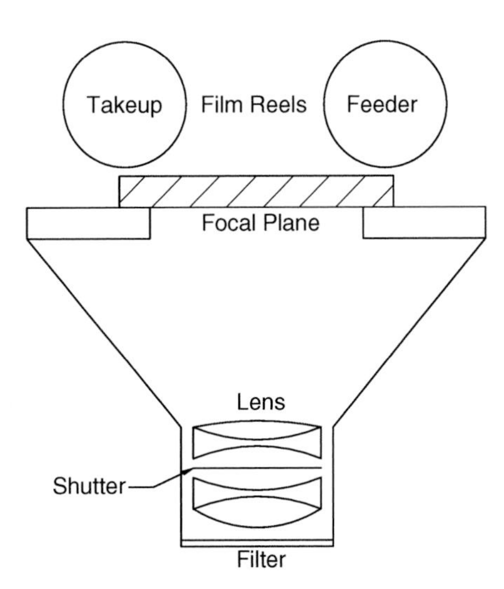

Figure 4.4 Major components of an analog aerial camera.

4.6.2.1 Magazine

A detachable magazine fits on top of the camera cone. It is separate from the frame of the camera and can be transported as a disengaged unit. This film magazine contains the aerial film supply and takeup spools as well as the pressure plate for flattening the film. Also included is a built-in forward image motion compensation device which eliminates photo image motion in the forward direction of the line of flight.

4.6.2.2 Film Reels

Contained within the magazine are two film reels. The feeder reel contains unexposed film, while the takeup reel holds the exposed film. After the exposure is made, a motor pulls the exposed frame onto the takeup reel. At the same time, a length of unexposed film is pulled from the feeder reel.

4.6.2.3 Focal Plane

A frame of unexposed film lies along the underside of the focal plane before the exposure is made. A vacuum is applied to the film at the instant of exposure so that the film is held flat against the focal plane. Otherwise air bubbles would collect beneath the film, causing uncontrollable distortions on the photographic image.

4.6.2.4 Lens Cone

The lens cone is a stable framework which separates the lens assembly from the focal plane. The lens cone contains a compound lens element between the lens shutter and motorized drivers. The lens system is compound, meaning that there are several elements of polished glass.

Camera shutters are constructed as a series of thin metal plates that overlap one another. The shutter mechanism is located between the front and rear elements of the lens system. When the shutter is activated, these plates slide open to form an aperture which admits light to expose the film. This opening is known as the diaphragm. The photographer uses f-stop/speed exposure setting combinations to control the amount of admitted light and length of time that the film is exposed to light.

The length of time that the shutter remains open is contingent upon factors that allow the entrance of sufficient light to make a normal exposure. Shutter speeds can vary through an interval ranging from $^1/_{60}$ to $^1/_{1000}$ of a second.

4.6.2.5 Image Motion

A photographic exposure is not instantaneous. Rather, when an exposure is made the shutter is open for some period of time, which varies with the type of film and the amount of radiant light. During this interval the camera is in motion and is subjected to movements.

The camera moves forward as the aircraft advances along the flight path. Since the film is stationary within the camera, this movement of the camera platform during an exposure period can cause blurring of the image. This is especially true during periods of marginal lighting when the shutter remains open for longer time spans. Additionally, engine vibrations are conducted throughout the airframe, and cameras fitted into a solid mount respond to the influence of this movement.

Camera mounting systems that reduce the effects of these types of vibrations and subsequent image blur are used in most aerial image platforms. These systems are commonly referred to as forward motion compensation (FMC) mounting systems.

4.6.3 Camera System

An analog aerial camera system is a complex arrangement of interrelated, high-technology, electromechanical accessories. Refer to Chapter 4 in *Aerial Mapping Methods and Applications* (Lewis Publishers, Boca Raton, FL, 1995) to view a schematic showing the essential components of a "state-of-the-art" precision aerial camera system.

4.6.4 Focal Length

Imbedded within the rear lens elements is a point, measurable with optical calibration equipment, known as the rear nodal point. The focal length of a given camera is the finitely measured distance from the rear nodal point within the lens to the focal plane (see Figure 4.5). Focal length is important to the planner in that it is a function of flight height determination.

There are several common focal lengths available: narrow angle (12 in.), normal angle (8.25 in.), wide angle (6 in.), and superwide angle (3.5 in.). Image analysis projects may utilize all of these various focal lengths, whereas photomapping projects use 6 in. predominantly and 3.5 in. to a limited extent.

4.6.5 Camera Calibration Report

Periodically, every two or three years, precision aerial mapping cameras are submitted to the U.S. Geological Survey (USGS) to be tested. This procedure

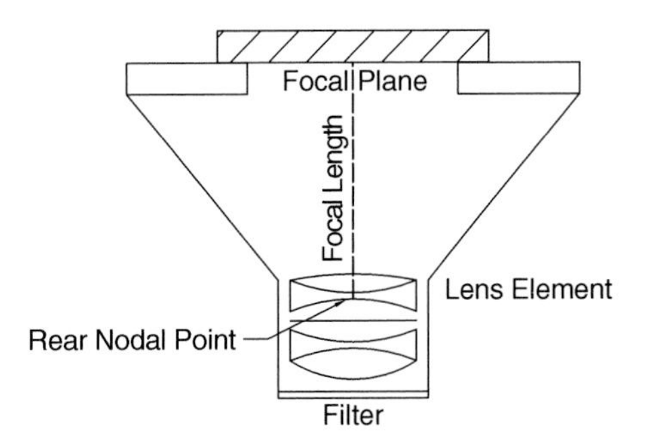

Figure 4.5 Focal length of an aerial camera.

precisely measures a number of operating procedures to assure that the camera consistently functions in an approved manner. The USGS inspector provides the owner with a comprehensive report detailing the results of the calibration test procedures. Some of the items enumerated in the report are input into the control matrix of the stereoplotter to aid in stereomodel orientation.

4.6.6 Digital Camera Components*

Digital camera systems are relatively similar to analog systems. These digital imagery systems can collect black and white, natural color, and color infrared imagery. The option of having natural color and color infrared imagery collected from a fixed winged or rotary winged aircraft can be very advantageous. Environmental assessment often requires imagery data collection at a specific and often narrow band of time. Digital imagery such as that shown in Color figures 1 and 2** allowed for the collection of both natural color and color infrared video of a site in Alaska. Digital imagery such as this may be incorporated into a softcopy workstation immediately after collection. No intermediate steps such as film processing, prints, or film positives are required prior to data manipulation and analysis. The most significant differences between analog and digital camera systems are the charge-coupled device (CCD) and the digital image storage device. Digital camera systems do not utilize photographic film to record an image. Rather, they utilize a CCD to record an image as a matrix of pixels along with a computer data storage device to record a group of image data sets (images). Figure 4.6 shows a schematic of a digital camera system.

The CCD can vary in storage capacity and resolution, which affects the clarity of the digital image. Clarity and resolution of a digital camera image is improved by capturing the imagery at lower altitudes. However, lower altitudes require more image files for a specific project ground area and more image storage capacity. Recent advances in CCD development, computer processing, and data storage capacities are beginning to make digital camera systems competitive with analog systems for many projects.

Those who wish to gain further information about digital cameras may benefit by accessing the keyword phrase "digital aerial camera" on the Internet. Several specific sites may be beneficial:

- http://Possys.com/
- http://.www.ziimaging.com/Products/AerialCameraSystems/RMK_Top.htm
- http://www.1h-systems.com/photogrammetry.htm

A discussion on ADAR multispectral cameras found at http://www.geo.wvu.edu./geog455/spring98/01/intro.htm may also be of interest.

* Aerospace Corporation offers a primer on the airborne digital camera at the web site http:/www.aero.org/publications/GPSPRIMER/GPSElements.htm.
** Color figures follow page 42.

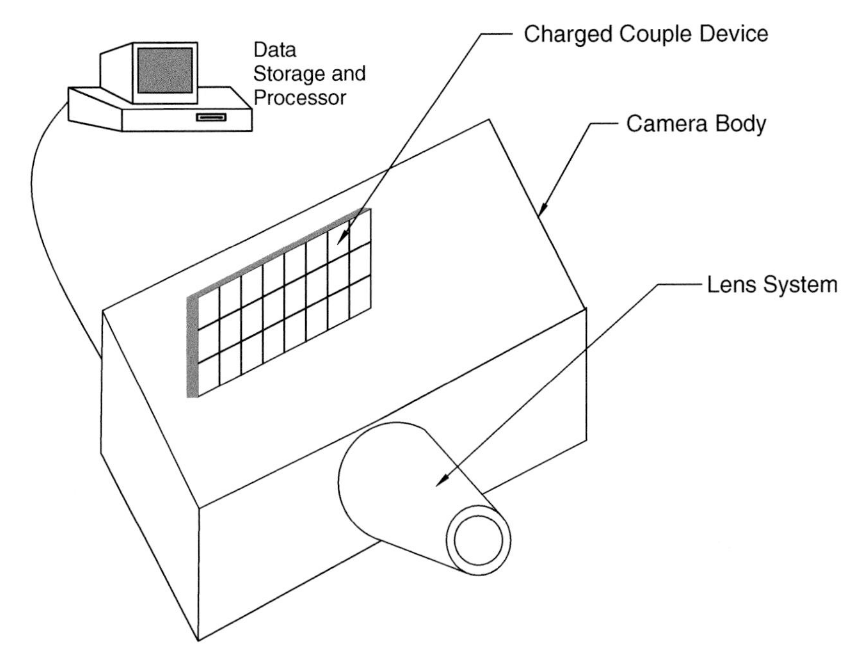

Figure 4.6 Schematic of a digital camera.

CHAPTER 5

Aerial Photographs

5.1 NOMENCLATURE OF AN AERIAL PHOTOGRAPH

Precision aerial mapping cameras most commonly furnish a photographic format with an image area 9 × 9 in. plus a 0.5-in. border on all four sides. Refer to the aerial photograph in Chapter 6, Figure 6.2.*

5.2 USES OF AERIAL PHOTOGRAPHS

Aerial photographs can be utilized in both planimetric and topographic mapping as well as in photo interpretation or image analysis for various disciplines. Sometimes a single type of film is best for a particular use, but for some applications several film types can be used in combination.

Aerial photography utilizing color films is somewhat more expensive than black and white film; however, in some situations the additional cost may be overshadowed by the amount of extra detail that can be extracted from one film type as opposed to another. A list of some uses that take advantage of various film types is found in Appendix II of *Aerial Mapping: Methods and Applications* (Lewis Publishers, Boca Raton, FL, 1995).

5.3 TIME-LAPSE PHOTOGRAPHY

There are many applications for time-lapse air photo comparison, where the aerial photos can be exposed over the same features periodically to detect changes during the interim period. A few examples are:

- Glacial movement
- Rock slide movement

* For a discussion of exposure information and fiducial marks refer to *Aerial Mapping: Methods and Applications,* Lewis Publishers, Boca Raton, FL, 1995.

- Long-term construction sites
- Advancing vegetation pathogens
- Timber cutting and reforestation tracts
- Wetlands expansion or decline
- Charting urban growth
- Monitoring flood pools
- Monitoring erosion
- Assessing disaster damage

5.4 SOURCES OF AERIAL PHOTOGRAPHS

Aerial photography for mapping and geographic information system (GIS) projects may be flown specifically for each project. Otherwise, the photo scale may not be compatible with the map scale and/or vertical mensuration. Image analysts can sometimes locate existing coverage for a project, since there are a number of sources from which to purchase aerial photographs. When dealing with these organizations, customers must accept whatever coverage is available, which may or may not suit the needs of the project. Those listed herein do not represent the complete market.

When requesting information from these organizations, it is advisable to provide them with a map specifically outlining the area of interest.

5.4.1. Private-Sector Mappers

Some mapping companies, especially those located near large metropolitan areas, periodically offer speculation aerial photography for sale to the general public. This is often valuable for historical study, business location forecasting, or trend analysis. These firms are also available for custom photo projects.

In addition, their film libraries contain negatives from clients who may give permission for reproduction products. There are various sources for locating firms which engage in aerial photographic activities. The inquirer can reference the local phone book, or inquiries can be made through the long distance information operators in appropriate regions.

5.4.1.1 Management Association for Private Photogrammetric Surveyors

The Management Association for Private Photogrammetric Surveyors (MAPPS) is an organization of private-sector aerial mapping companies in the United States and several foreign countries. It may be contacted through:

Management Association for Private Photogrammetric Surveyors
12020 Sunrise Valley Drive
Suite 100
Reston, VA 22091

Their web site http://www.mapps.org/ lists its member companies alphabetically by name and by individual states in the United States or in Canada, Germany, India,

Poland, and New Zealand for easy access to pertinent information such as address, contacts, telephone, Fax, and e-mail. Web sites of many firms can be directly referenced to gain some insight to their organizations, facilities, and services.

5.4.1.2 American Society for Photogrammetry and Remote Sensing

ASPRS is a professional society whose membership is comprised of remote sensing scientists, photogrammetrists, GIS users, and others who rely on these disciplines. Numerous mapping companies and firms in related endeavors are also sustaining members of this society. The ASPRS publishes a monthly technical journal, *Photogrammetric Engineering and Remote Sensing.* Information may be requested from:

American Society for Photogrammetry and Remote Sensing
5410 Grosvenor Lane, Suite 210
Bethesda, MD 20814-2160
301-493-0290 (Tel)
301-493-0208 (Fax)
http://www.asprs.org

5.4.2 Federal

There are a number of federal agencies which can furnish copies of existing aerial photographic coverage.

5.4.2.1 Agricultural Stabilization and Conservation Service

All counties in the U.S. are photographed periodically, every ten years or so, under the ongoing Agricultural Stabilization and Conservation Service (ASCS) program. Each county agent retains a set of the most recent coverage. These are available for perusal within the office of the agent. Copies of photos from previous flights may be ordered from:

U.S. Department of Agriculture
Agricultural Stabilization and Conservation Agency
Aerial Photography Office
2222 West 2300 South
P.O. Box 30010
Salt Lake City, UT 84130-0010

5.4.2.2 U.S. Geological Surveys

This agency maintains a massive ongoing mapping and map revision program throughout the United States. Their library files contain volumes of aerial photo coverage throughout the country. Information can be obtained from:

Earth Sciences Information Center
U.S. Geological Survey
1400 Independence Road
Rolla, MO 65401

5.4.2.3 National Archives

An enormous amount of historical federal aerial photography prior to 1950 is available for purchase from:

Cartographic Branch (NNSC)
National Archives
Washington, D.C. 20408

5.4.2.4 The Earth Resources Observation Satellite

The Earth Resources Observation Satellite (EROS) Data Center is a repository for resource imagery. Information can be supplied by:

Data Services Branch
EROS Data Center
Sioux Falls, SD 57198

5.4.2.5 Others

Many federal agencies maintain files of aerial photographs for in-house use. Unless restricted for security purposes, reproductions of these may be purchased from the agency which exposed the photographs. The U.S. Army Corps of Engineers, U.S. Forest Service, Bureau of Reclamation, Bureau of Land Management, U.S. Fish & Wildlife, or Bureau of Indian Affairs may be viable sources. Information about these coverages should be directed toward the appropriate regional, division, district, or facility office.

5.4.3 States

Individual states maintain agencies that require aerial photographs. Information may be acquired from the appropriate state department of transportation or department of natural resources.

5.4.4 Counties and Municipalities

Most county tax assessors have a store of aerial photographs, and many municipalities maintain files of photographs. Queries should be directed toward the appropriate county or municipal assessors or city managers.

CHAPTER 6

Geometry of Aerial Photographs*

6.1 SCALE EXPRESSIONS

Map or photo scale can be stated in one of two expressions as applied to aerial mapping: representative fraction or engineers' scale.

6.1.1 Representative Fraction

Representative fraction is expressed as a ratio in the form of 1:2400, where one unit on the photo or map represents 2400 similar units on the ground. For example:

1 in. on the map/photo = 2400 in. on the ground
1 ft on the map/photo = 2400 ft on the ground
1 m on the map/photo = 2400 m on the ground

6.1.2 Engineers' Scale

Engineers' scale is expressed as a ratio in the form of 1 in. = 200 ft, where one unit on the photo or map represents a number of different units on the ground.

6.1.3 Scale Conversion

Both of the examples of scale cited above mean the same thing. In order to convert from a representative fraction of 1:2400, assume that both are in inches. Then 2400 in. divided by 12 in. equals 200 ft, so the resultant engineers' scale is 1 in. = 200 ft. On the photo or map, 1 in. is equal to 200 ft on the ground.

Conversely, to convert from an engineers' scale of 1 in. = 200 ft, multiply 200 ft by 12 in. This simple arithmetic exercise equates 1 in. on the map or photograph to 2400 in. on the ground. The resultant representative fraction would be 1:2400.

* The geometry discussed in this chapter is reduced to several simple formulae that can be easily utilized in planning aerial photographic missions.

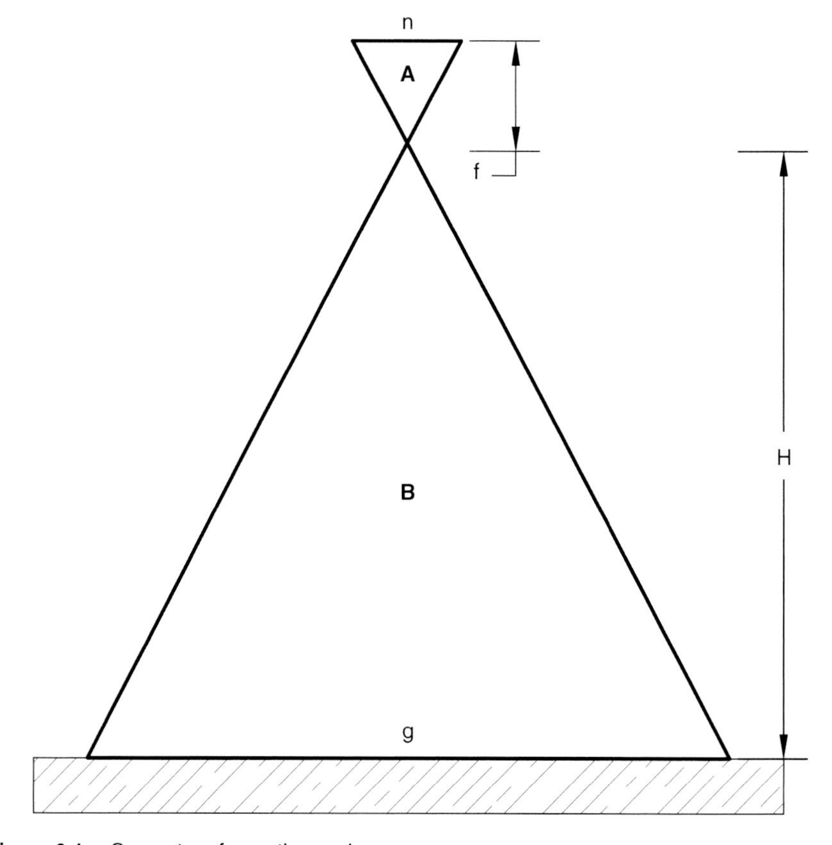

Figure 6.1 Geometry of negative scale.

6.2 GEOMETRY OF PHOTO SCALE

Figure 6.1 identifies similar triangles **A** and **B**. Analogous parts of similar triangles are proportional. Therefore, **n** (negative width) is proportional to **g** (ground distance covered by exposure) in the same magnitude as **f** (focal length) is to **H** (flight height above mean ground level).

6.2.1 Derivation of Photo Scale

The derivation of photo scale (s_p) with Equation 6.1 is a ratio which serves the purpose of determining negative scale based on negative width related to ground distance covered by the exposure frame.

$$s_p = \frac{n}{g} : \frac{1 \text{ in.}}{x \text{ ft}} \tag{6.1}$$

By factoring this ratio, the engineers' scale is 1 in. = x ft.

6.2.2 Controlling Photo Scale

Negative scale can be controlled by considering the specific mission characteristics. Negative scale is computed with the aid of Equation 6.2. This defines the relationship between focal length and flight height above mean ground level.

$$s_p = \frac{f}{H} : \frac{1\text{ in.}}{x\text{ ft}} \tag{6.2}$$

By factoring the ratio, the photo scale is 1 in. = x ft.

6.2.2.1 Engineers' Scale

When using a 6-in. focal length camera flying at a height of 1200 ft above mean ground elevation, the engineers' scale of the photograph is:

$$s_p = \frac{f}{H} : \frac{6\text{ in.}}{1200\text{ ft}} = \frac{1\text{ in.}}{200\text{ ft}}$$

which translates to an engineers' scale of 1 in. = 200 ft.

6.2.2.2 Representative Fraction

This same situation can be utilized to calculate representative fraction, keeping in mind that a 6-in. focal length is equal to 0.5 ft:

$$s_p = \frac{f}{H} = \frac{0.5\text{ ft}}{1200\text{ ft}} = \frac{1\text{ ft}}{2400\text{ ft}}$$

which is analogous to a representative fraction of 1:2400.

6.2.3 Scale Formula

A simplified formula for determining photo scale, derived from the photo scale ratio, is presented in Equation 6.3.

$$s_p = H/f \tag{6.3}$$

where:

s_p = photo scale denominator (feet)
H = flight height above mean ground level (feet)
f = focal length of camera (inches)

Figure 6.2 Large-scale aerial photograph. (Courtesy of Surdex Corporation, Chesterfield, MO.)

6.2.4 Flight Height

For photos exposed with most precision aerial-mapping cameras, the calibrated focal length is noted in the margin of the exposure. A derivation of Equation 6.3 is Equation 6.4 for calculating flight height once the photo scale is selected.

$$H = s_p * f \tag{6.4}$$

6.2.5 Relative Photo Scales

There can be some confusion when thinking about relative photo scales. Just remember, large scale means that image detail is relatively large, and small scale means that image detail is relatively small. The discernable village that is visible on the photograph in Figure 6.2 appears on a large-scale aerial photo. Compare this with the photograph in Figure 6.3, which contains the same village in a small-scale aerial photo.

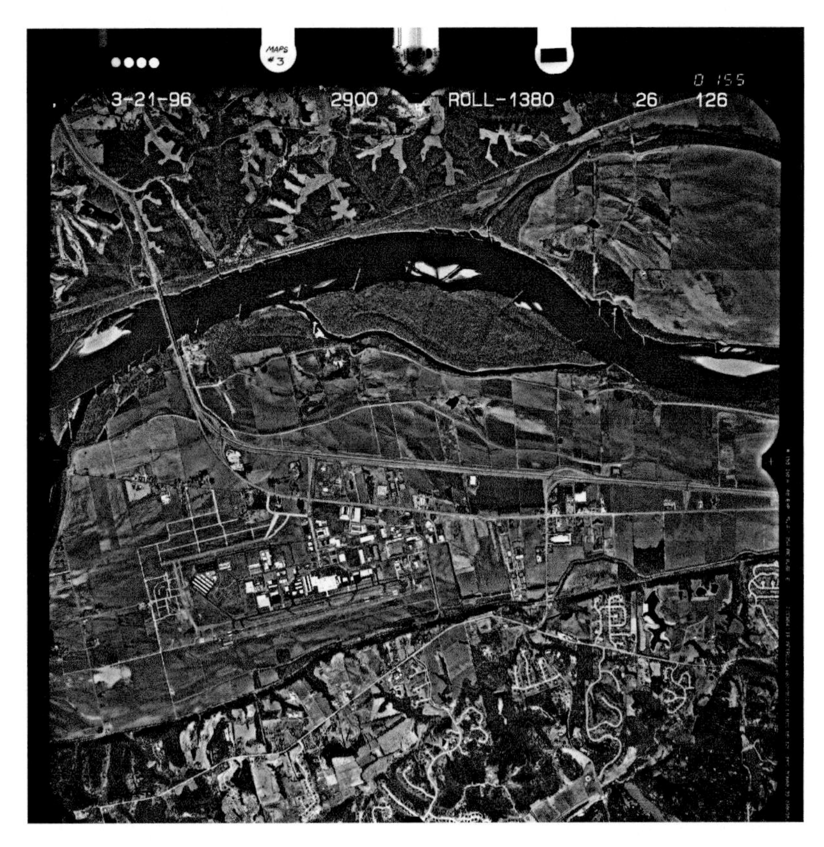

Figure 6.3 Small-scale aerial photograph. (Courtesy of Surdex Corporation, Chesterfield, MO.)

6.3 PHOTO OVERLAP

Aerial photo projects for all mapping and most image analyses require that a series of exposures be made along each of the multiple flight lines. To guarantee stereoscopic coverage throughout the site, the photographs must overlap in two directions: in the line of flight and between adjacent flights.

6.3.1 Endlap

Endlap, also known as forward overlap, is the common image area on consecutive photographs along a flight strip. This overlapping portion of two successive aerial photos, which creates the three-dimensional effect necessary for mapping, is known as a stereomodel or more commonly as a "model." Figure 6.4 shows the endlap area on a single pair of consecutive photos in a flight line.

Practically all projects require more than a single pair of photographs. Usually, the aircraft follows a predetermined flight line as the camera exposes successive overlapping images.

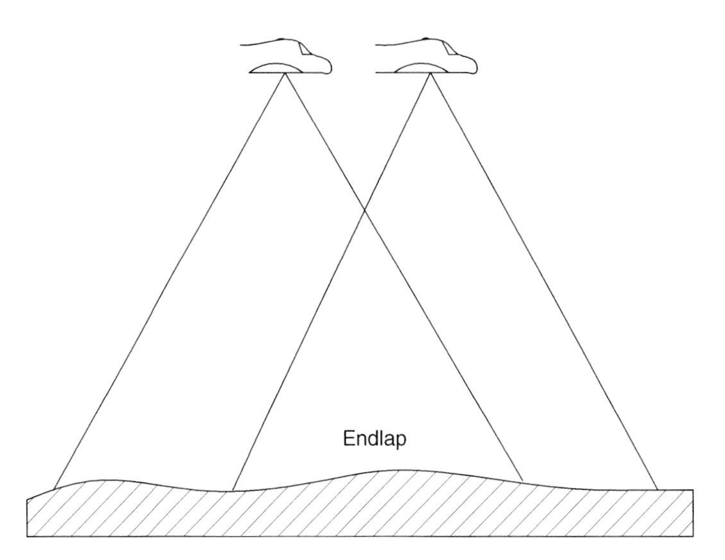

Figure 6.4 Endlap on two consecutive photos in a flight line.

Normally, endlap ranges between 55 and 65% of the length of a photo, with a nominal average of 60% for most mapping projects. Endlap gain, the distance between the centers of consecutive photographs along a flight path, can be calculated by using Equation 6.5.

$$g_{end} = s_p * w * \left[\left(100 - o_{end}\right)/100\right] \qquad (6.5)$$

where:

g_{end} = distance between exposure stations (feet)
s_p = photo scale denominator (feet)
o_{end} = endlap (percent)
w = width of exposure frame (inches)

When employing a precision aerial mapping camera with a 9 × 9 in. exposure format and a normal endlap of 60%, the formula is simpler. In this situation, two of the variables then become constants:

w = 9 in.
o_{end} = 60%

Then, the expression $w*[(100\% - o_{end})/100]$ becomes a constant equal to 3.6, and Equation 6.5 may be supplanted by Equation 6.6.

$$q_{end} = s_p * 3.6 \qquad (6.6)$$

When utilizing a camera other than a 9 × 9 in. format and/or an endlap other than 60%, Equation 6.5 must be employed.

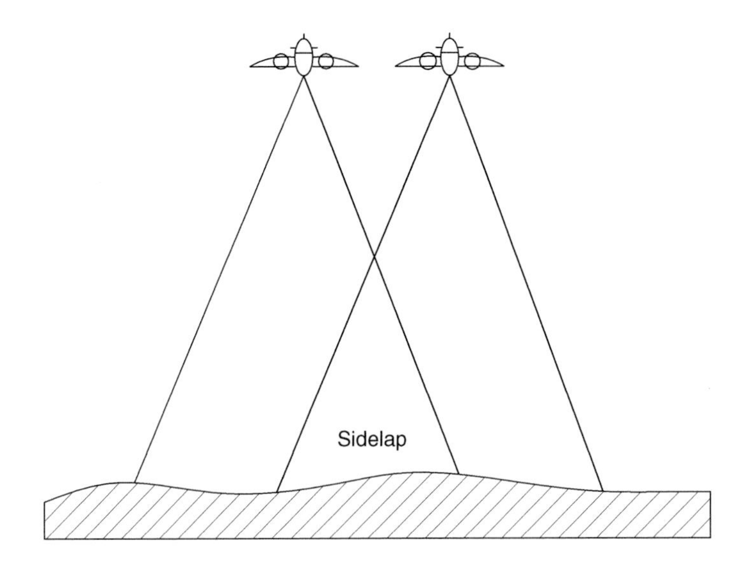

Figure 6.5 Sidelap between two adjacent flight lines.

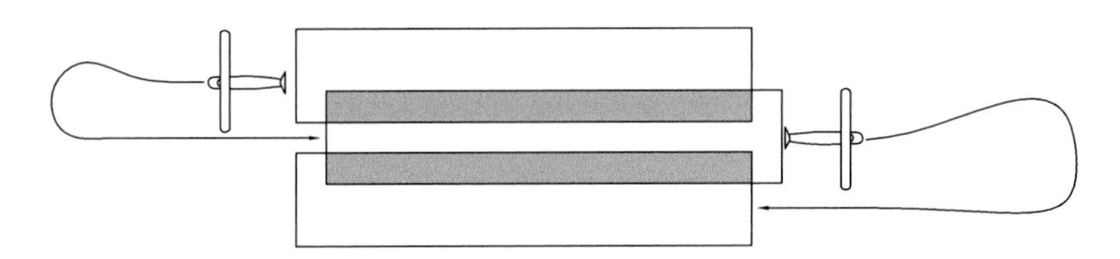

Figure 6.6 Sidelap on three adjacent flight lines.

6.3.2 Sidelap

Sidelap, sometimes called side overlap, encompasses the overlapping areas of photographs between adjacent flight lines. It is designed so that there are no gaps in the three-dimensional coverage of a multiline project. Figure 6.5 shows the relative head-on position of the aircraft in adjacent flight lines and the resultant area of exposure coverage.

Usually, sidelap ranges between 20 and 40% of the width of a photo, with a nominal average of 30%. Figure 6.6 portrays the sidelap pattern in a project requiring three flight lines.

Sidelap gain, the distance between the centers of adjacent flight lines, can be calculated by using Equation 6.7.

$$q_{side} = s_p * w * \left[\left(100 - o_{side}\right)/100\right] \tag{6.7}$$

where:

g_{side} = distance between flight line centers (feet)
s_p = photo scale denominator (feet)
o_{side} = sidelap (percent)
w = width of exposure frame (inches)

When employing a precision aerial mapping camera with a 9 × 9 in. exposure format and a normal sidelap of 30%, the formula is simpler. In this situation, two of the variables then become constants:

w = 9 in.
o_{side} = 30%

Then, the expression w*[(100% – o_{side})/100] becomes a constant equal to 6.3, and Equation 6.7 may be supplanted by Equation 6.8.

$$q_{side} = s_p * 6.3 \tag{6.8}$$

When utilizing a camera other than a 9 × 9 in. format and/or a sidelap other than 30%, Equation 6.7 must be employed.

6.4 STEREOMODEL

From the foregoing discussion of overlap, it is evident that consecutive photos in a flight strip overlap. When focusing each eye on a particular image feature that was viewed by the camera from two different aspects, the mind of the observer is convinced that it is seeing a lone object with three dimensions. Put simply, the three-dimensional effect is an optical illusion. This phenomenon of observing a feature from different positions is known as the parallax effect. Although used to describe other facets of photogrammetry, parallax is defined as a change in the position of the observer. This situation allows a viewer, when using appropriate stereoscopic instruments, to observe a pair of two-dimensional photos and see a single three-dimensional image.

Photogrammetrists envision a model as the "neat" area that a single stereopair contributes to the total project. This allows for the endlap and sidelap with surrounding photos. A mapping model is shown as the crosshatched area in Figure 6.7.

Table 6.1 is a tabulation relating photo scale to flight height using a camera with a 6-in. focal length. For a given photograph, several parameters can be found:

- Flight height (above mean ground level)
- Photo center interval
- Flight line spacing
- Acres per model (neat area)

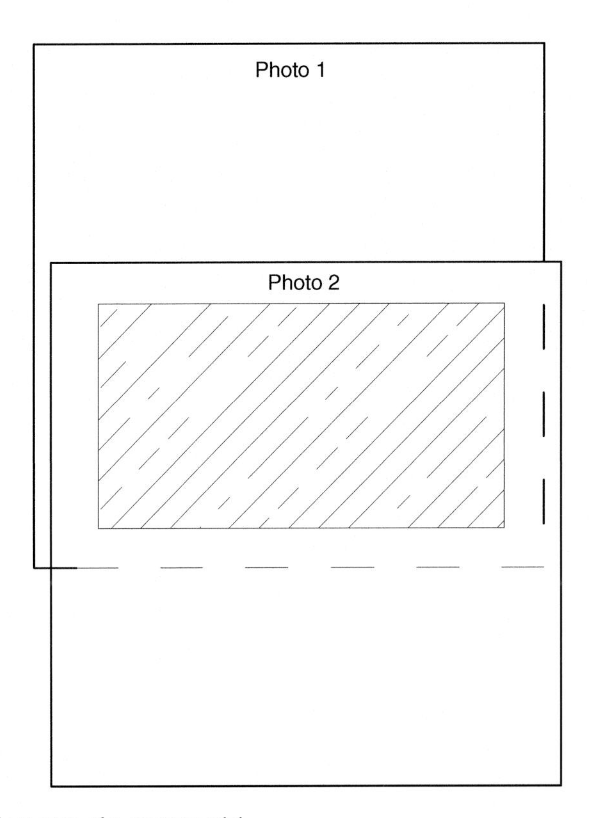

Figure 6.7 Neat area of a stereomodel.

It must be realized that the scale of individual photographs in a project is not a constant. Due to undulations in the aircraft flight and terrain relief, the distance between the camera and the ground differs from one exposure to another. Therefore, photo scale must be considered as an average scale for the total project.

6.5 RELIEF DISPLACEMENT

The surface of the earth is not smooth and flat. As a consequence, there is a natural phenomenon that disrupts true orthogonality of photo image features. In this respect, an orthogonal image is one in which the displacement has been removed, and all of the image features lie in their true horizontal relationship.

6.5.1 Causes of Displacement

Camera tilt, earth curvature, and terrain relief all contribute to shifting photo image features away from true geographic location. Camera tilt is greatly reduced or perhaps eliminated by gyroscopically-controlled cameras.

Table 6.1 Tabulation of Photo Scales with the Resultant Flight Height (Above Mean Ground), Endlap Gain, Sidelap Gain, and Acreage Per Neat Model

Scale 1 in. = *x* in.	Flight Height	Gain/Photo	Gain/Line	Acres/Model
167	1,000	601	1,052	14
200	1,200	720	1,260	21
250	1,500	900	1,575	32
300	1,800	1,080	1,890	47
350	2,100	1,260	2,205	64
400	2,400	1,440	2,520	83
450	2,700	1,620	2,835	105
500	3,000	1,800	3,150	130
550	3,300	1,980	3,465	158
600	3,600	2,160	3,780	187
650	3,900	2,340	4,095	220
700	4,200	2,520	4,415	255
750	4,500	2,700	4,725	293
800	4,800	2,880	5,040	333
850	5,100	3,060	5,355	376
900	5,400	3,240	5,670	422
1,000	6,000	3,600	6,300	521
1,250	7,500	4,500	7,875	813
1,320	7,920	4,753	8,316	907
1,500	9,000	5,400	9,450	1,171
1,667	10,000	6,000	10,500	1,446
2,000	12,000	7,200	12,600	2,983
2,500	15,000	9,000	15,750	3,245
3,000	18,000	10,800	18,900	4,685

Earth curvature is of little consequence on large-scale photography. The relatively small amount of lateral distance covered by the exposure frame introduces only a minimal amount of curvature, if any.

Topographic relief can have a great effect on displacing image features. The amount of image displacement increases on high-degree slopes. Feature displacement also increases radially away from the photo center.

6.5.2 Effects of Displacement

An aerial photograph is a three-dimensional scene transferred onto a two-dimensional plane. Hence, the photographic process literally squashes a three-dimensional feature onto a plane that lacks a vertical dimension, and image features above or below mean ground level are displaced from their true horizontal location. Figure 6.8 illustrates this phenomenon. Assuming that the stack rises straight into the air from the ground, both the top and the base possess the same horizontal (XY) placement. This diagram belies that fact, because the base and the top are in displaced positions (labeled “d” in Figure 6.8) on the negatives. This separation will not be of the same magnitude on successive photos.

Figure 6.9 illustrates the radial displacement of an object in an aerial photograph.

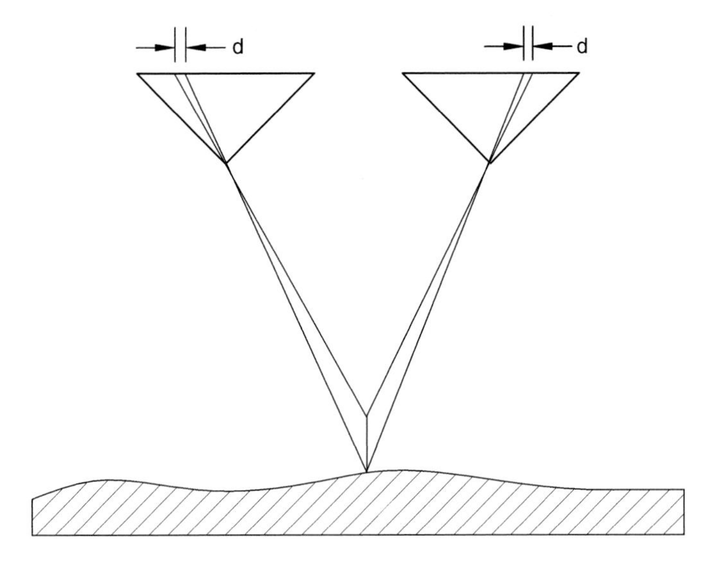

Figure 6.8 Image displacement.

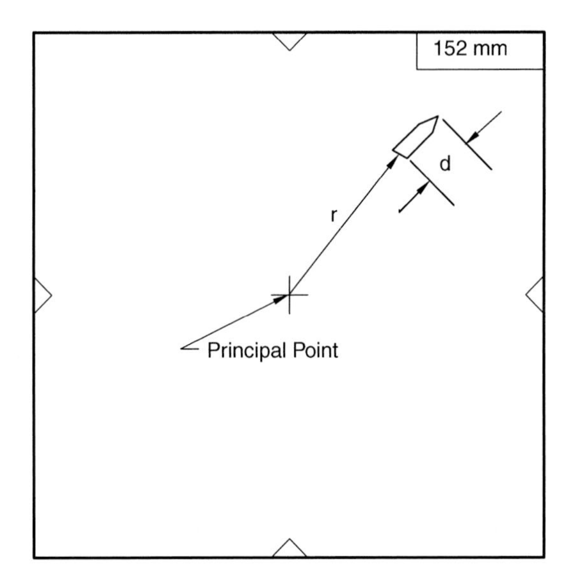

Figure 6.9 Radial displacement of an image feature.

Just as images of fast-rising features are displaced, so are the changes in ground elevations, though not as visibly apparent in the photographs. Figure 6.10 illustrates relief displacement on a straight utility clearing that crosses rolling hills. The clearing is identified as the wavy open strip running diagonally through the woods on the left side of the photo. Even though the indicated utility clearing follows a straight course, relief displacement due to terrain undulations causes this feature to waver.

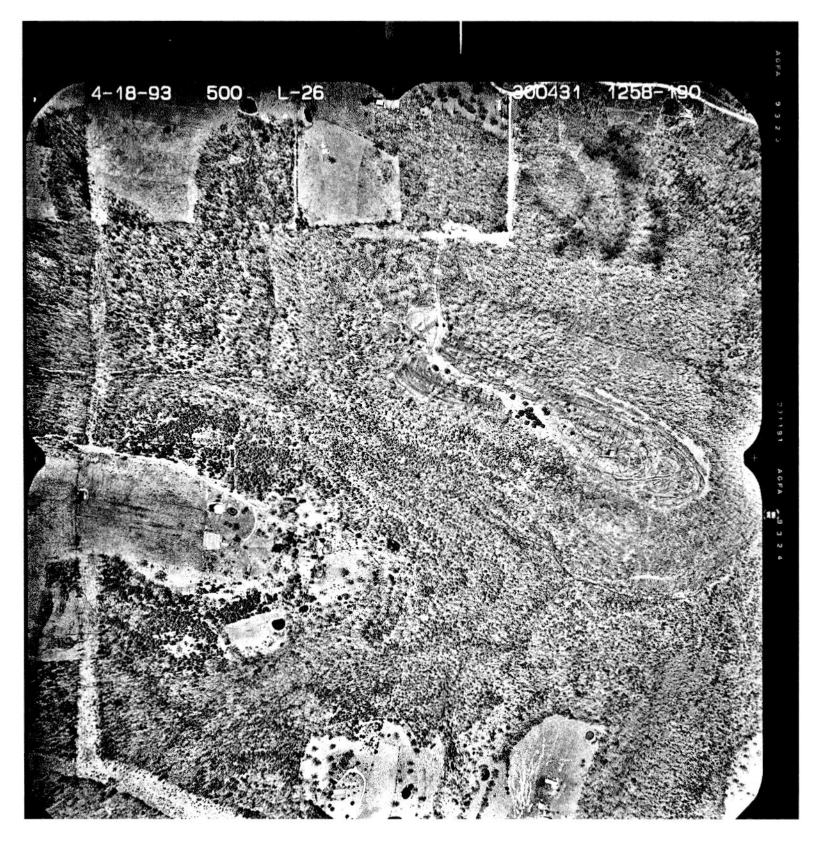

Figure 6.10 Image displacement on a utility clearing. (Courtesy of Surdex Corporation, Chesterfield, MO.)

Presuming this to be true, it follows that if several scale sets are calculated from an individual photograph, each may vary from the others. So, the more diverse the terrain character is, the more the scale variance.

6.5.3 Distortion vs. Displacement

Often, the term distortion is considered to be synonymous with displacement.

Distortion implies aberration. It is caused by discrepancies in the photographic, processing, and reproduction systems. This condition is not correctable in the compilation of a stereomodel.

Displacement is a normal inherent condition. Since mapping instruments work with a three-dimensional spatial image formed by a pair of overlapping two-dimensional photos, predictable displacement can be compensated for in the mapping process. Rather than being a fault in the image structure, displacement is the means by which it is possible to extract spatial information from photographs.

6.6 MEASURING OBJECT HEIGHT

Relief displacement allows the measurement of image object heights, either from a single photo or from a stereopair. Although photo interpreters in the past put these procedures to good use by manual methods, contemporary mappers are not directly concerned with the implementation of this approach because mapping instruments rely upon analytical solutions employing higher mathematics to achieve greater accuracy in processing differential parallax* comparator readings to create the same solution.

Differential parallax is an important concept in photogrammetric mapping. It allows the coordination of map features from images. Essentially, softcopy mappers and digital stereoplotters rely upon differential parallax to accomplish digital data collection.

* For a basic study of differential parallax refer to Chapter 6 in *Aerial Mapping: Methods and Applications,* Lewis Publishers, Boca Raton, FL, 1995.

Color Figure 1

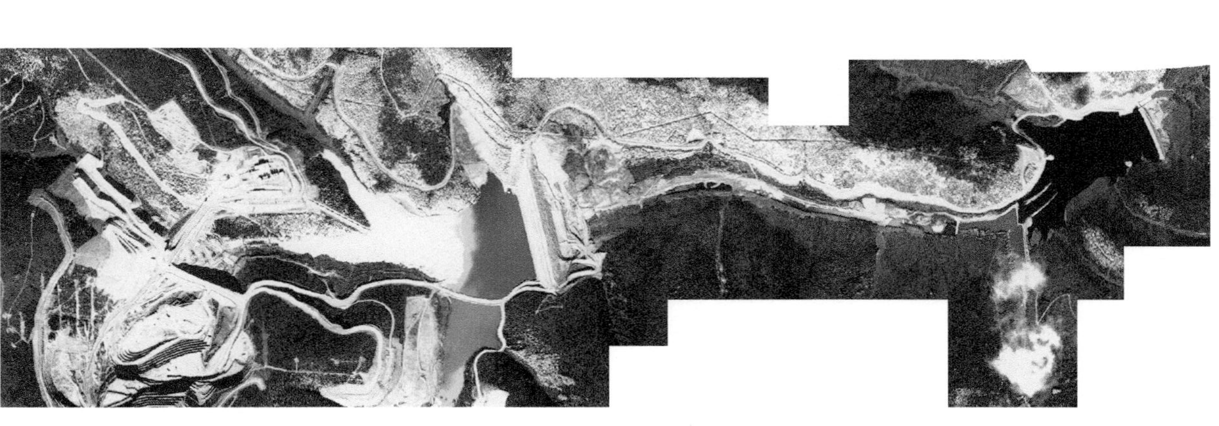

Color Figure 2

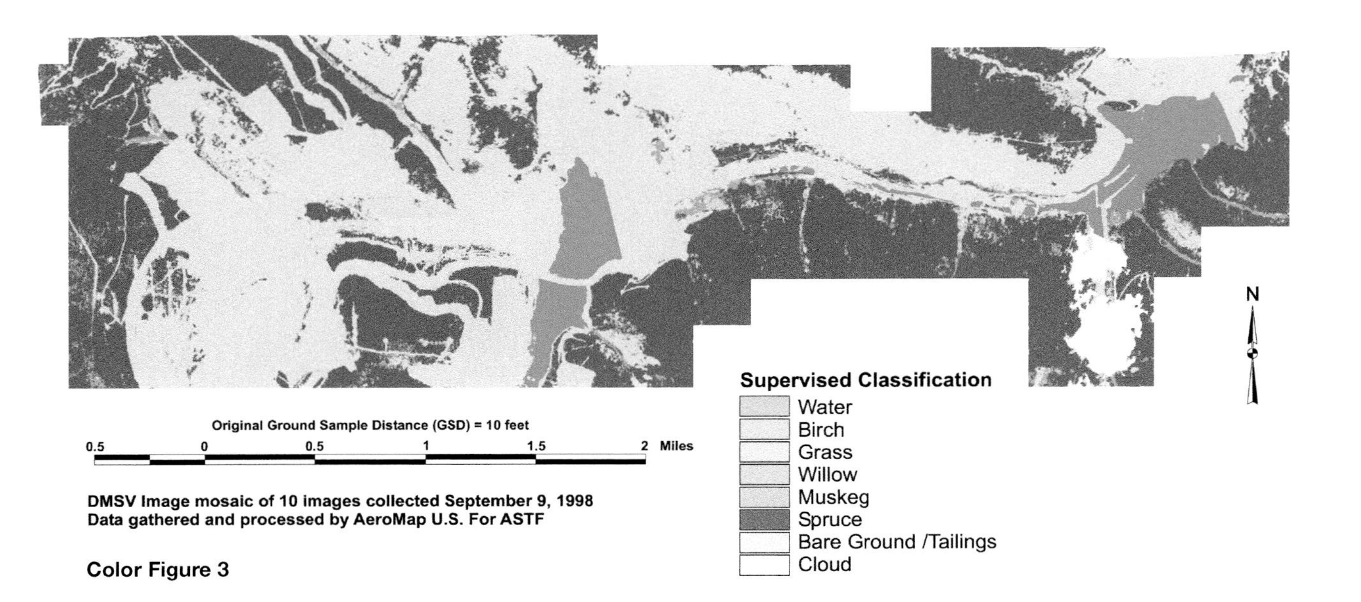

Color Figure 3

Color Figure 1. Natural color imagery. This imagery is part of an Alaska Science and Technology Foundation grant, "Remote Sensing Technology for Mining Applications". John Ellis, of AeroMap, was the Project Manager. Courtesy of AeroMap U.S., Anchorage, AK, with permission.

Color Figure 2. False color imagery. This imagery is part of an Alaska Science and Technology Foundation grant, "Remote Sensing Technology for Mining Applications". John Ellis, of AeroMap, was the Project Manager. Courtesy of AeroMap U.S., Anchorage, AK, with permission.

Color Figure 3. A thematic map created by supervised classification procedures. This imagery is part of an Alaska Science and Technology Foundation grant, "Remote Sensing Technology for Mining Applications". John Ellis, of AeroMap, was the Project Manager. Courtesy of AeroMap U.S., Anchorage, AK, with permission.

CHAPTER 7

Map Compilation

7.1 HISTORY

Producing accurate commercial maps from aerial photography began in the 1930s. The technology of stereomapping over the last 70 years has brought vast technological improvements.

7.1.1 Stereoplotters

Stereoplotters are machines that incorporate complex optical train viewing systems with computer-driven precision instrumentation. These components are employed in extracting, analyzing, and recording spatial information from aerial photographs. A spatial model is a stereopair of photographs referenced to its true geographic position, and stereoplotters utilize models in the collection of spatial data. A pair of near-vertical aerial photographs exposed from two different locations, with sufficient overlap of the area of interest, allows the operator to view a two-dimensional image in three dimensions. Stereoplotters use film positives of the exposures of interest as the spatial stereopair media. These film positives are commonly referred to as diapositives in the photogrammetic mapping industry and are produced as a second-generation positive transparency on machines, such as that seen in Figure 7.1, which electronically dodges the image to eliminate undesirable light and dark areas.

The operator inserts consecutive overlapping diapositives into receptacles residing within the stereoplotter. Even though the compiler is actually viewing two separate images, proper relative alignment of the photos allows the operator's mind to fuse images into a reduced three-dimensional diorama of the overlapping area as it appears to float in spatial limbo.

Figure 7.2 is a view of an analytical stereoplotter in operation, similar to a selection of instruments marketed by various manufacturers that are the mainstay of contemporary stereomapping. Items that are visible in Figure 7.2 are:

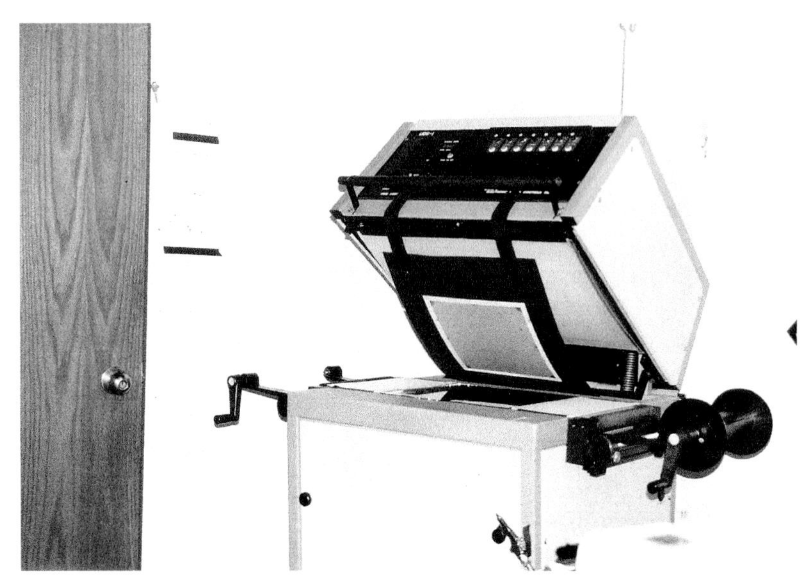

Figure 7.1 An electronic dodging machine used to produce film plates employed in stereo-mapping. (Photo courtesy of authors at Walker and Associates, Fenton, MO.)

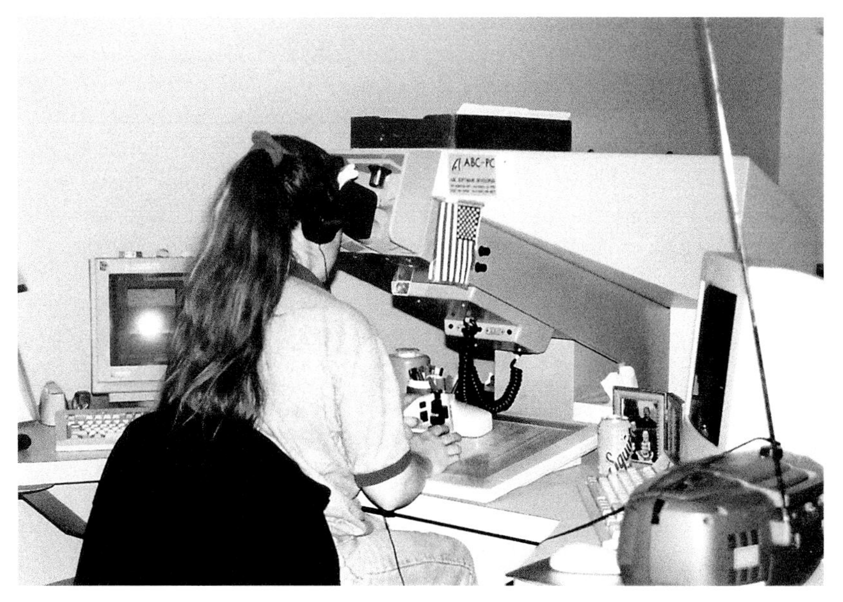

Figure 7.2 An analytical stereoplotter in operation. (Photo courtesy of authors at Walker and Associates, Fenton, MO.)

- The monitor to the left of the technician controls the computer which drives the system.
- The monitor to the right of the technician displays the map compilation features.
- The binoculars allow the technician to view the three-dimensional image.
- The tracing cursor allows the technician to guide the internal floating reference mark to compile the mapping data.

7.1.1.1 Georeferencing

To compile spatial data in their correct location, both horizontally and vertically, the stereomodel must be referenced to a prescribed vertical and horizontal datum. In the surveying and mapping industry this is referred to as georeferencing the stereomodel. A minimal number of ground points that are identifiable in the stereopair must be known for proper georeferencing. The operator employs prescribed procedures to measure and adjust the location of stereopairs in space to the locations of the known ground features. These procedures are repeated until an acceptable spatial location of the stereopair is obtained and satisfactorily georeferenced.

After the photos are properly oriented, the internal reference mark is maneuvered horizontally and vertically to read the true XYZ (east/north/elevation) geographic position of any point on the stereomodel.

7.1.1.2 Data Compilation

Once a stereopair is properly georeferenced in a stereoplotter, the data can be viewed, interpreted, and compiled. Data compilation is the collection of information pertaining to and describing features within a stereomodel, accomplished by interpretation of the object type whether it is topography, structures, utilities, natural features, or vegetation. Data compilation may also include the shape and position (horizontal and vertical) of a selected feature relative to other features in the stereopair. Topography is compiled as points of known elevation, which can be mathematically interpreted into contours (lines of equal elevation) by software applications.

7.1.2 Evolution of Stereoplotters

The implementation of stereoplotters has passed through three distinct mechanical generations and is now in a digital environment. Mechanical stereoplotters can be generally defined by the fact that they require film products of the area of interest and an optical train (a series of lens and prisms) device to view in three dimensions. Digital stereoplotting systems are commonly referred to as softcopy mapping systems, which incorporate high-density scanned images, coupled with high-resolution computer screens, high-speed processors, unique suites of map feature compilation software, and digital data storage devices.

Readers who wish to learn more about the obsolete mechanical forerunners of contemporary mapping instruments can refer to Chapter 7 in *Aerial Mapping: Methods and Applications* (Lewis Publishers, Boca Raton, FL, 1995).

7.1.2.1 Digital Stereoplotters

The 1980s ushered in the third generation of stereoplotter, which is equipped with internal control computer systems and data collectors. These systems, termed analytical, create a mathematical image solution, allowing the computer to perform model orientation. High-speed computers and complex software are employed as

Figure 7.3 A softcopy system. (Courtesy of Surdex Corporation, Chesterfield, MO.)

drivers that move the data collection point to the desired precise location for feature collection.

The optical/mechanical construction of a digital instrument allows for the collection of feature coordinates simultaneously from two successive overlapping photographs. This improved feature position accuracy is accomplished by computer software that interprets and converts the differential parallax of feature coordinate sets from the overlapping photographs into precise geographic locations.

When the map compilation is in progress, some analytical systems require the operator to work from a three-dimensional image while looking directly into the viewing binoculars, and the compiled data appear on a graphic screen. A combination software/hardware unit interjects the generated digital information into the optical path of the train of optics and reproduces it directly on top of the spatial photo image.

*7.1.2.2 Softcopy Systems**

In the 1990s computer processing speeds increased dramatically, almost simultaneously with continuous increases in data storage. The technological advancements in computer processing and data storage also drove the demand for very high-resolution image scanners. These technology changes and demands drove the mapping industry to a total softcopy solution for mapping.

* The internet keywords "softcopy mapping" leads to several pertinent web sites pertaining to softcopy mapping, including references to equipment, images, contour maps, orthophotos, digital elevation models, GIS, and other applicable items of interest.

The operation of a softcopy system such as that seen in Figure 7.3 requires a different approach from that employed in the operation of stereoplotters. Each of the photos used to form the stereomodel is scanned, and the data are input into a database.

Similar to the first analog systems of the 1960s and 1970s, each photo in a stereopair is viewed independently by the operator with the use of polarized glasses, enabling the technician to view each image, correct for parallax, and view the overlapping area of the stereopair in three dimensions.

The web site http://www.rwell.com/dms.htm describes a desktop mapping system that operates on the softcopy principle.

Softcopy mapping systems generally require several components, including a large format graphic screen, a high-speed central processing unit, large data storage units, and a high-resolution metric scanner, coupled with several software packages. Software packages would be required to perform georeferencing of stereopairs, digitizing, annotation, image manipulation, and database management. Additional software packages may be required to perform GIS functions.

Recent developments in digital camera systems allow mapping system developers to eliminate the film and diapositives. Overlapping digital images of a mapping area are collected and stored as digital data with the use of high-resolution metric digital cameras. This digital image data is then entered directly into the softcopy mapping system. As the technology of digital camera imagery approaches the image quality of silver-grained emulsion film, total digital photogrammetry is becoming a reality. When properly employed, digital camera imagery is capable of producing the same mapping products and required accuracies as those produced with traditional film products. The time needed to complete mapping projects could also be reduced by the elimination of film processing, diapositive production, and printing.

7.1.3 Future Developments

Photogrammetric mapping system technology undergoes constant evolution due to the constant changes in computer technology. The need and demand for geographic information also drive these developments. There is an ever-increasing need to pinpoint happenings in the world. Digital camera technology enhancements include multispectral and hyperspectral camera systems in light fixed- and rotary-winged aircraft. Laser mapping technology — light detection and ranging (LIDAR) — has developed to a point where it is being deployed in fixed and rotary aircraft to collect accurate point locations (X, Y, and elevation), which can then be processed to produce topographic information. Satellite mapping system development is not only performed by major governments; private industry around the world will be developing orbiting mapping systems with data available to the private sector at competitive rates. Mapping systems will be developed that eliminate the need to manually collect many common features such as buildings, roads, and vegetation. Mapping production technology will increase the compilation speed and accuracy to a point where it is collected in near real time. These changes are ongoing, and many of them have been under development for several years.

7.2 DATA FORMAT

Information is created and stored in several different modes.

7.2.1 Raster Format

Raster data are scanned from a photograph or graphic image or are collected by a multispectral scanner, thermal scanner, or radar. The image is composed of rows and columns of numerous picture elements, individually known as pixels. Each pixel has a defined size and specific incident flux value indicating a specific type of information. Radiation intensity is graded from 0 (no reflectance), which exhibits a dark image, to 255 (full reflectance), which exhibits a bright image. Figure 7.4 illustrates a set of raster data aligned in columns and rows of picture elements with a single pixel isolated as the hatched frame.

Many mapping projects require the mapping scientist to scan imagery in order to incorporate it into a mapping system for feature compilation and/or analysis, which may require a high-resolution metric scanner. Figure 7.5 illustrates a scanner in operation. A photo scanner is basically a densitometer working as an analog-to-digital converter. It records the radiometric density value of a discrete pixel on the photographic image in the form of a binary integer.

High-resolution photo scanners are generally constructed as flatbed or rotary and are capable of scanning translucent as well as opaque media. Scanners used for map compilation today may also be capable of scanning color imagery. Some color scanners scan red, green, and blue in separate passes, while others are capable of scanning all three primary colors in one pass. Scanning devices are designed to hold the film in place as a light source passes over it, and a photomultiplier records the image density of each pixel. Some of the major differences in scanners include the flexibility of scan resolutions, the scanners maximum and minimum scan resolution, one-pass or three-pass color scanning, and the ability to scan from a roll of film. Most of these differences, with the exception of scan resolution capability, only affect the time to produce a scan and not the quality.

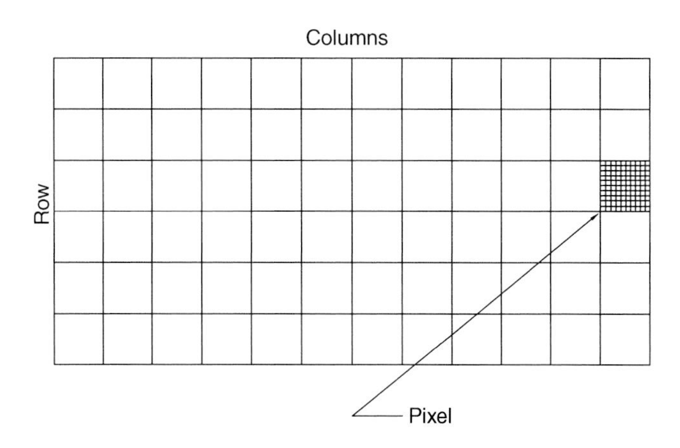

Figure 7.4 Raster data format.

Figure 7.5 A scanner in operation. (Photo courtesy of authors at Walker and Associates, Fenton, MO.)

Pixels extracted from the raster scan of a photograph could be as small as 1/1000 in. or even as minute as 1/5000 in. for use in some softcopy environments. The ground dimension of pixels collected by resource satellites is in the range of 10 or 20 m from the French Systeme Pour L'Observation de la Terre (SPOT) satellites and 30 m or 120 m from the American LANDSAT resource satellite. The resolution of pixels captured with meteorological satellite technology is even larger, exceeding 1 km.

7.2.2 Vector Format

Vector data are generated in planar or spatial Cartesian coordinate data strings, as illustrated in Figure 7.6, forming points, lines, and polygons:

- A point is located by a single XY or XYZ coordinate set.
- Lines are composed of a series of points connected by vectors.
- A polygon is a succession of lines enclosing an area.

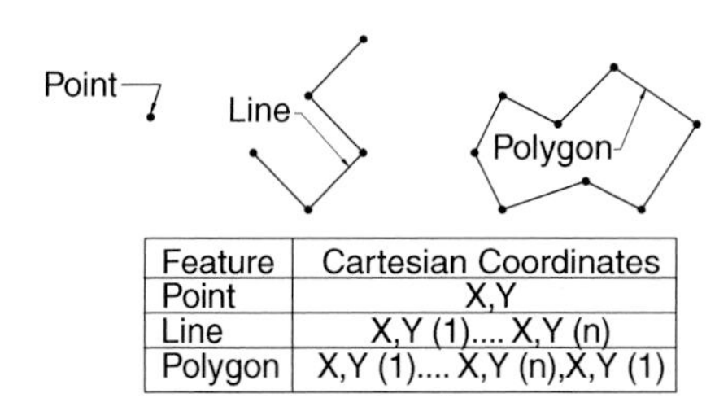

Feature	Cartesian Coordinates
Point	X,Y
Line	X,Y (1).... X,Y (n)
Polygon	X,Y (1).... X,Y (n),X,Y (1)

Figure 7.6 Vector data format.

Figure 7.7 A data editing system. (Photo courtesy of authors at Walker and Associates, Fenton, MO.)

7.2.3 Attributes

Collateral textual information resides in the database to describe, explain, supplement, or identify the raster and vector data. Tabular attribute information can be retained in accessible matrices.

7.3 DIGITAL OUTPUT

Once the data are collected, it is input into a data editing system such as the one shown in Figure 7.7 to assure its reliability.

Output from map compilation and geographic analysis can include the following:

- Natural and man-made cultural features
- Digital terrain model data, breaklines, and spot elevations for the development of contours, wireframes, perspective views, and orthophotos
- Terrain profile data to be used in stockpile or excavation volumes and earth movement calculations
- Digital orthorectified photographs

Output media for map compilation data may include magnetic tapes, hard drives, diskettes, and CDs. Hardcopy data sets may include line maps or image renditions, or a composite of both can be created from the raster or vector files on a data plotter. Hardcopy media may include bond paper, velum, and/or photographic paper plots.

CHAPTER 8

Map Accuracies

8.1 QUALITY ASSURANCE/QUALITY CONTROL

For the first several decades of mapping from aerial photos utilizing stereoplotters, users could physically see and handle graphical mapping products. Now, the preponderance of end products from a mapping project resides, unseen by the human eye, as a matrix of magnetic spots within an electronic database. Therefore, the user must rely upon the integrity of captured information to maintain the credibility of a mapping endeavor.

8.1.1 Significance of Quality Assurance/Quality Control

The significance of quality assurance and quality control (QA/QC) in contemporary photogrammetry cannot be overly emphasized. QA/QC is an integral factor in the data collection task and cannot be sacrificed to expediency or false economy.

A relatively limited QA/QC effort is usually directed toward field tests to verify the quality of final mapping products. As a consequence, credible judgment must be exercised by both mapper and user in designing a mapping project so that the accuracy of the final output adheres to reliable tolerances.

Quality control should be the watchword of the mapper who is obligated to institute professional quality control procedures. Quality assurance should be ascertained by the user who is committed to performing quality assurance assessment measures.

8.1.2 Funding vs. Quality

In the initiation of mapping projects, the user asks two questions:

- How much will it cost?
- How soon will it be completed?

There is no denying that more stringent mapping requirements are more costly. If the user's funding is insufficient to purchase mapping that will adhere to required accuracy standards, the project should not be undertaken. A judicious reevaluation of user needs may prove that a smaller map scale and/or a larger contour interval would satisfy the needs of the project. In this situation the specifications, but not the accuracy standards, could legitimately be revised to provide a lower cost.

Users must accept the fact that products from a lower accuracy level cannot be electronically manipulated to fulfill the requisites of a higher level product. No amount of computer manipulation will increase the accuracy level of a block of data. Reliable mappers do not diminish standards to provide a lower cost.

8.2 RAMIFICATIONS OF FAULTY MAPPING

Unrealistic product delivery schedules can initiate serious accuracy dilemmas. There can be no denying that human stress resulting from pressures applied to expedite all phases of mapping production promotes mistakes. Faulty mapping can and often does result in very serious consequences.

8.2.1 Rework

If errors are detected during data capture, the mapping can be revised. These errors can originate in the field surveys, aerotriangulation, or compilation stage. Rework may be precipitated also by a change in specifications during the ongoing course of the project. Regardless of the cause, rework is a "triple-pronged" cost penalty:

1. Original cost. Initially, the user must pay the cost to have the original mapping produced.
2. Out of pocket. Either the user or responsible contractor (surveyor or photomapper), depending upon where the fault lies, is faced with the out-of-pocket expense to accomplish corrected mapping.
3. Lost revenue. If responsible for the corrective work, the surveyor or photomapper forgoes any revenue which could be realized doing income-producing work for others during that same period of rework production time.

8.2.2 Abandoned Schedules

When corrections are necessary, the elapsed time could very well be twice the original production schedule. The condition that precipitates the need for rework can occur during any phase of the mapping procedure.

A problem encountered at any point during the course of the production scheme may necessitate additional effort in previous phases. Rework causes irretrievable slippage in the production schedule.

8.2.2.1 Original Schedule

A certain amount of time is required to accomplish the various sequential phases to complete the original mapping. Aerial photographers, surveyors, and photomappers must schedule the work of multiple users and accomplish the work for each on a scheduled rotation.

8.2.2.2 Revised Schedule

If rework is required in any phase, the schedules of all of the users in pursuant phases must be revised to meet this contingency. The time consumed to complete revisions will vary, depending upon the extent of the effort required to accomplish the rework.

8.2.3 Design Failures

In the event that mapping corrections are not detected until well into the use of the maps, there could be design aberrations built into the project for which the mapping was obtained. This may lead to wasted effort on the part of the user or costly design changes during development of the project.

8.2.4 Legal Action

Any of the above factors can and increasingly does lead to costly lawsuits and court action. Both time and money are valuable. If one or the other is needlessly squandered through neglect on the part of the designer, surveyor, or mapper, legal efforts for punitive damage compensation may be initiated.

8.3 MAP ACCURACY STANDARDS

Relevance of an accuracy concept cannot be overly stressed. The design of a mapping project should be guided primarily to assure adherence to the intended accuracy of the end product. Upon these specifications rests the utility of a mapping product.

Uppermost in the minds of both mapper and user should be the precept that if the collected digital data does not meet accuracy standards, the end product is of little value. In fact, inaccurate mapping creates a negative impact in that it encourages a false sense of reliance upon inferior information.

8.3.1 Various Map Accuracy Standards

A number of map accuracy standards are used by those who produce and use maps. Some have been developed for use within individual agencies, and others have been created for a wider range of users. Agencies that may use individual specifications include the Federal Emergency Management Agency (FEMA), U.S. Department of Transportation, U.S. National Cartographic Standards for Spatial Accuracy, and various state and local governments. The deluge of digital data sets being produced by various federal agencies has helped them to come together in a common group known as the Federal Geographic Data Committee (FGDC) and to develop the most recent standards, known as the Geospatial Positioning Accuracy Standards. These standards are also known as the National Standard for Spatial Data Accuracy (NSSDA) and can be reviewed in detail by logging on to the www.fdgc.gov/standards web site.

Although many standards are available, only National Map Accuracy Standards (NMAS), ASPRS Standards, and FGDC NSSDA (July 1998) will be discussed herein. The three standards are nationally recognized to encompass most current spatial data accuracy requirements. NMAS and ASPRS standards are both referenced and discussed in the NSSDA standards.

All standards discussed herein allow for deviation from accuracy standards when the terrain surface on the photo image is obscured by clouds, shadows, or vegetative cover.

8.3.2 National Map Accuracy Standards (1947)

From 1941 to the mid-1990s, most map production organizations, both in public and private sectors, accepted NMAS as an industry standard for both large- and small-scale photogrammetric mapping. Prevailing procedures, techniques, and equipment called for in the NMAS provided predictable mapping products and accuracies. The NMAS were generally developed around mapping products developed by federal agencies such as the USGS and the U.S. Department of Agriculture (USDA). Accuracies specified in the NMAS are stated in terms of the position of a feature on a hardcopy map in relation to its spatial position on the earth. These features may include planimetric features such as buildings and roads, contours (lines of equal elevation), and individual spot elevations.

NMAS, in its elementary interpretation, stipulates levels of accuracy for both horizontal and vertical features:

- Planimetric features: 90% of finite cultural objects should be accurate to within 1/40 in., and 100% should be correct to within 1/20 in. at delivered map scale.
- Contours: 90% of contours should be accurate to within one-half a contour interval, and 100% should be correct to within one full contour interval.
- Spot elevations: 90% of plotted spot elevations should be accurate to within one-fourth of a contour interval, and 100% should be correct to within one-half a contour interval.

8.3.3 American Society for Photogrammetry and Remote Sensing (1990)

ASPRS standards requirements are developed around procedures and equipment used to produce large-scale maps. These standards establish the map accuracy by establishing the difference between the location of a feature in a spatial data set and its true position on the earth. This standard stipulates that the difference should be derived by making comparative measurements of the feature position on the earth using a more accurate method than was used in the photogrammetrically derived feature position (i.e., ground surveys). The difference is stated in terms of root mean square error (RMSE) between the photogrammetrically derived feature position and its corresponding position as measured by a more accurate means. ASPRS standards for large-scale mapping are divided into three classes. They are described as follows:

Class 1 is the most stringent.
Class 2 can contain inaccuracies twice that of Class 1.
Class 3 errors can be triple those of Class 1.

ASPRS standards establish a number of production parameters to confine errors within the enumerated limits. These tolerance levels are expressed as the RMSE of all of the test point inaccuracies encountered by field verification.

Discrepancy in coordinate direction at individual test points can be computed using Equation 8.1. This formula can be utilized for the X, Y, or Z coordinate error.

$$e = v_{map} - v_{test} \tag{8.1}$$

where:
e = X or Y or Z coordinate discrepancy
v_{map} = X or Y or Z map coordinate
v_{test} = X or Y or Z field survey coordinate

To determine the actual RMSE of X, Y, and Z coordinates, the sum of the square of the individual errors must be determined. This can be accomplished with Equation 8.2.

$$E^2 = \sum \left(e_1^2 + e_2^2 + \cdots + e_n^2\right) \tag{8.2}$$

where:
E^2 = sum of the square of the errors
Σ = sum
e_1^2 = coordinate discrepancy at first point
e_n = coordinate discrepancy at nth point

The actual RMSE of any tested group of map data, in ground feet, is calculated by using Equation 8.3.

$$e_{rms} = \sqrt{E^2/n} \tag{8.3}$$

where:
E_{rms} = actual RMSE
n = number of points tested

8.3.3.1 Horizontal Inaccuracies

Comparable horizontal inaccuracy allowances stated by ASPRS standards for ASPRS Class 1 mapping can be determined by using Equation 8.4.

$$e_{h1} = s_m/100 \tag{8.4}$$

where:
s_m = map scale denominator (feet)
e_{h1} = maximum allowable ASPRS Class 1 RMSE

Once the ASPRS Class 1 inaccuracy is calculated, the error tolerance for ASPRS Class 2 can be determined with Equation 8.5.

$$e_{h2} = e_{h1} * 2 \tag{8.5}$$

ASPRS Class 3 error can be determined with Equation 8.6.

$$e_{h3} = e_{h1} * 3 \tag{8.6}$$

8.3.3.2 Contour Inaccuracies

Comparable vertical inaccuracy allowances stated by ASPRS Class 1 standards, relative to contour interval, can be determined through the use of Equation 8.7.

$$e_{c1} = c_i/3 \tag{8.7}$$

where:
c_i = contour interval
e_{c1} = maximum allowable ASPRS Class 1 error

After computing ASPRS Class 1 allowable errors, Equation 8.8 can be used to determine the allowable error for ASPRS Class 2 contour deviation.

$$e_{c2} = e_{c1} * 2 \tag{8.8}$$

Equation 8.9 can be utilized to furnish anticipated contour deviation for ASPRS Class 3.

$$e_{c3} = e_{c1} * 3 \tag{8.9}$$

8.3.3.3 Spot Elevation Inaccuracies

Maximum allowable errors of spot elevations for ASPRS Class 1 can be determined by applying Equation 8.10.

$$e_{s1} = c_i/6 \tag{8.10}$$

where:

c_i = contour interval

e_{c1} = maximum allowable ASPRS Class 1 error

Once the ASPRS Class 1 allowable error is computed, the ASPRS Class 2 error can be determined by using Equation 8.11.

$$e_{s2} = e_{s1} * 2 \tag{8.11}$$

The ASPRS Class 3 error can be determined with Equation 8.12.

$$e_{s3} = e_{s1} * 3 \tag{8.12}$$

8.3.4 Federal Geographic Data Committee

The NSSDA was established by the FGDC, a group of federal agencies that produce and use geospatial data. A committee was established under the FGDC to update the current NMAS and ASPRS standards, taking into account new technology equipment. This group also recognized that since the 1990s, the demand for digital geospatial data has increased dramatically.

The users and their accuracy requirements are also very diverse. Although the NSSDA uses RMSE in the same manner as the ASPRS standard, the accuracy is reported as the acceptable RMSE at 95% confidence level. The NSSDA refers to and allows for reporting of accuracy in terms of NMAS or ASPRS standards. A spatial data user who wants to specify a spatial data set horizontal and/or vertical accuracy in NSSDA terms could choose an ASPRS Class or NMAS scale and associated accuracy and state it in terms of 95% confidence level (NSSDA). The web site www.fgdc.gov/standards provides detailed explanations and examples of NSSDA accuracy calculations, as well as conversion of NMAS and ASPRS accuracies to NSSDA.

8.3.4.1 Horizontal Standard Error

To calculate the horizontal standard error in terms of 95% confidence level according to NSSDA, log on to the web site www.fgdc.gov/standards and click on Part 1 to find the derivation of the following formulas:

Horizontal Accuracy

The horizontal accuracy calculation assumes one of two scenarios:

- When $RMSE_x = RMSE_y$

  LET $RMSE_r = sqrt(2*RMSE_x^2) = sqrt(2*RMSE_y^2)$, and

  $Accuracy_r = 1.7308*RMSE_r$

- When $RMSE_x \neq RMSE_y$,

  LET circular $Accuracy_r = \sim 1.2239*(RMSE_x + RMSE_y)$

Vertical Accuracy

$$\text{LET RMSE}_z = \text{sqrt}[\Sigma(z_{data\ i} - z_{check\ i})^2/n]$$

where:

$z_{data\ i}$ = elevation of ith check point in the data set
$z_{check\ i}$ = surveyed elevation of ith check point in the independent test
n = number of points being checked
i = integer from 1 to n

then,

$$\text{Accuracy}_z = 1.96 * \text{RMSE}_z$$

8.4 PROCEDURAL SUGGESTIONS

Although NMAS and ASPRS standards are not wholly compatible, both serve a meaningful purpose — to assure that both the map producer and the user are aware that it is essential to preserve product quality. A review of accepted NMAS and ASPRS standards would note that accuracies for a specified mapping horizontal map scale and vertical interval are different. Standards will indicate that, for practical purposes, NMAS inaccuracy allowances are bracketed by ASPRS Classes 1 and 2. NSSDA will allow for conversion of NMAS and ASPRS accuracies to a common accuracy, and it is recommended that this method of qualifying accuracy be considered when possible.

8.4.1 Cautions

There are reservations to be considered in recommending standards unreservedly by the tyro user, because the power of GIS queries can cause the inexperienced spatial data user to become enmeshed in accuracy problems. Mixing of data sets with diverse accuracies or accuracies stated in terms of different standards can be a recurring problem.

8.4.1.1 Historical Acceptance of National Map Accuracy Standards

For many decades NMAS were the predominant guidelines within the photogrammetric community for large-scale and small-scale mapping. For more than ten years there has been a concentrated effort among organized photogrammetrists to sponsor an attitude of stricter accuracy demands for large-scale mapping. Inaccuracy tolerances predicated by NMAS fall immediately below those in ASPRS Class 1, which is in direct agreement with this philosophy of improved quality.

8.4.1.2 Indiscriminate Data Use

There is a tendency, when indiscriminately collecting information to create a consolidated database, for the user to ignore the perils inherent in dissimilar error tolerances incorporated within those data. To the uninitiated, the quality of all of the information within a database can be of equal accuracy. This is a precarious assumption.

Table 8.1 Minimum Accuracy Levels to Employ by Purpose of the Map

Map Class	Purpose of Map
ASPRS1	Final design
	Earthwork calculations
	Volume of pits or piles
NMAS	Route location
	Preliminary design
	Project planning
	Rough terrain
	General planning

8.4.2 Options

With this groundwork in place, two options are suggested.

8.4.2.1 Experienced User

In the hands of an experienced user who thoroughly understands disparate data quality and has complete in-house control over the use of the data, the various classes within the ASPRS mapping standards can be applied with confidence. The user may also decide to state the accuracy in NSSDA terms. This will allow the use of disparate data sets with a clear understanding of the accuracy of each in terms of RMSE at 95% confidence level. This instance presumes that the knowledgeable user is fully aware of allowable errors and their consequential deformations upon the final product.

8.4.2.2 Inexperienced User

In keeping with the doctrine of looking to maintain increased accuracy, users not fully aware of the pitfalls of integrating various accuracy sets may wish to combine the best of NMAS and ASPRS mapping standards. Even though the inaccuracy tolerances for all the ASPRS map classes will be listed in various tables throughout the book, it is recommended that the user limit choices to those suggested in Table 8.1.

After determining the inaccuracies allowed by ASPRS2 and ASPRS3, the user may wish to consider using these standards for mapping projects which are sufficiently tolerant of the magnitude of the errors that they may be capable of generating.

8.5 MERGING DIVERSE DATA

Information systems are a valuable tool in many fields of endeavor, but there are photogrammetric pitfalls in merging data gathered from diverse sources. The greatest hazard may stem from the ability of a computer driven by proper software to accept almost any matrix of digital XYZ data and to create a map to any scale or contour interval. Once data are collected from a variety of sources and assimilated into a single information system database, a tendency to treat all of the information similarly exists. Herein lies the fallacy. All features go into a database as a group of individual coordinate points which are relational to each other through a common geographic positioning grid. However, not all information is collected to the same degree of accuracy. A map is only as reliable as its most inaccurate information layer. Serious thought must be given to the compatibility of information that resides in an integrated database.

8.6 MAPPING SYSTEM ERRORS

Discounting the errors that have already been discussed, there are systemic aberrations in the photogrammetric functions in aerial mapping projects.

8.6.1 Photography

No aerial camera focal plane is absolutely flat. Certain areas of high and low spots and also inherent aberrations in lens systems exist. When using a camera that is calibrated periodically, these are normally not significant problems.

Due to the undulating character of the ground and the varying aircraft height, photograph scale fluctuates within a single exposure as well as between adjacent frames. This situation is adjusted during stereomodel orientation. Normal inherent relief displacement is compensated for in the sterocompilation procedure.

8.6.2 Stereocompilation

Diverse factors are error sources during digital data collection. Some can be attributed to human frailties, while others are dependent on outside influences.

8.6.2.1 Visual Acuity

Visual acuity varies with each photogrammetric technician. In some situations this may affect the vertical map accuracy by an amount approaching as much as one-fifth to one-fourth of a contour interval.

8.6.2.2 Image Definition

Resolution of the photo image may affect the operator's ability to place the reference mark on the true elevation of the image object. This could be a function of weather, film processing, or dispositive processing. Clear, crisp days tend to produce a "hard" or distinct image, and warm, humid days result in a "soft" or hazy image. Hard models allow the reference mark to be placed more precisely in contact with the terrain than soft models.

CHAPTER 9

Photo Scale Selection

9.1 CONTOUR FACTOR

Traditionally, American mappers have subscribed to a concept of a contour factor, which describes the geometric relationship between aircraft height (above mean ground level) and the smallest accurate contour interval that can be generated at a specific flight height. This correlation is referred to as the C-factor.

9.1.1 Application of the C-Factor

The C-factor is empirical rather than statistical. Hence its application is flexible, and professional acumen must be exercised in selecting the contour factor. The C-factor, as it is applied today in the United States, can be expressed by Equation 9.1.

$$c_f = H/c_i \tag{9.1}$$

where:

c_f = C-factor
c_i = contour interval (feet)
H = flight height (feet) above mean ground level

After choosing an appropriate contour interval (Table 9.1) and judging an equitable C-factor (Table 9.2), Equation 9.2 may be utilized to determine the flight height above mean ground level.

$$H = c_f * c_i \tag{9.2}$$

9.1.2 Influences upon C-Factor

Acceptable C-factors vary from one stereocompilation machine manufacturer and model to another. Actually, the C-factor for various mapping instrumentation is

Table 9.1 Recommended C-Factors to Achieve Specific Map Accuracy Standards

Contour Interval (ft)	Photogrammetric Mapping Purpose
1.0	Final design, earthwork computations, volumes
2.0	Route locations, preliminary design
4.0–5.0	Preliminary project planning
10.0	Steep terrain, general planning

Note: These C-Factors also apply to FGDC NSSDA.

Table 9.2 Maximum ASPRS and Interpolated NMAS C-Factors

	Stereomapping System	
Map Accuracy Standard	Softcopy Workstation	Analytical Stereoplotter
ASPRS Class 1	1600	2000
NMAS	1700	2100
ASPRS Class 2	1800	2200
ASPRS Class 3	2000	2500

considered as a range rather than a discrete integer. This range is influenced by many factors. The actual C-factor is project- and equipment-specific and can only be known after the project is completed and the mapping data are assessed for accuracy. Degradation of any, or the cumulative effect of several, of these elements will alter the precision of the C-factor. Refer to Chapter 9 in *Aerial Mapping: Methods and Applications* (Lewis Publishers, Boca Raton, FL, 1995) for an in-depth presentation of the specific production-oriented variables which influence its selection. Current technology advances such as digital cameras, softcopy workstations, and airborne GPS (ABGPS) also may affect the C-factor for a specific mapping project.

9.2 PHOTO SCALE/MAP SCALE/CONTOUR INTERVAL

Definitive geometrical relationships exist between map scale and contour interval, which determine the photo scale. Judicious implementation of these geometrical considerations greatly influences mapping integrity.

Selection of a reliable photograph scale is of major importance, because the quality of the final digital mapping product hinges primarily upon it. Three fundamental factors influence the selection of a photo scale for digital mapping:

- Equipment and system used in production
- Accuracy of the horizontal map scale
- Accuracy of the contour interval

It is mandatory that map scale and contour interval be considered separately prior to selecting a photo scale. Equipment and production systems hardware and software may affect both horizontal map scale and contour interval, as well as the overall selected photo scale.

9.2.1 Planimetric Features

On large-scale mapping projects, a great number of finite cultural features are compiled. These include, but are not limited to, poles, street signs, inlets, traffic signs, sidewalks, and manholes. As the map scale gets smaller, the end user may choose to omit some of the finite detail. The reason for this is that the features may not be visible and/or identifiable on the photos or that their exclusion would reduce map clutter and digital file size and performance. Some of the smaller features may be symbolized due to minimum size limitations. This dictates that large-scale planimetric mapping requires large-scale photos.

Current emphasis on GIS applications of spatial data may dictate that a high level of detail be captured and stored in a spatial data set. Detailed GIS query demands may become the driving force for final map scale and contour interval. An engineering design-based mapping project may have multiple uses to include incorporation into a facilities management GIS. The GIS may demand utility locations at a horizontal accuracy that would be greater than that required for general engineering purposes. In these situations it may be prudent to design the photo scale for a compilation level that will accommodate both engineering and GIS demands.

Past practice would discourage these GIS demands, but technology advancements are helping to accommodate them. Computer hardware (data storage devices and processors) is constantly getting faster and more capable of handling larger amounts of data reliably at relatively lower cost. Data compression routines and software are available that allow for reliable file size compression to 20 times reduction. Data storage media such as CDs and removable hard drives are increasing storage capability at reduced cost.

9.2.2 Photo Scale/Map Scale

To preserve horizontal validity of planimetric detail, the enlargement from the photographic image to the map should not exceed those factors listed in Table 9.3, which also apply to the Federal Geographic Data Committee (FGDC) National Standard for Spatial Data Accuracy (NSSDA) specifications.

Table 9.3 Maximum Recommended Enlargement Factors from Photo Scale to Map Scale

	Stereomapping System	
Map Accuracy Standard	Softcopy Workstation	Analytical Stereoplotter
ASPRS Class 1	6.0	7.0
NMAS	6.5	7.5
ASPRS Class 2	7.0	8.0
ASPRS Class 3	8.0	9.0

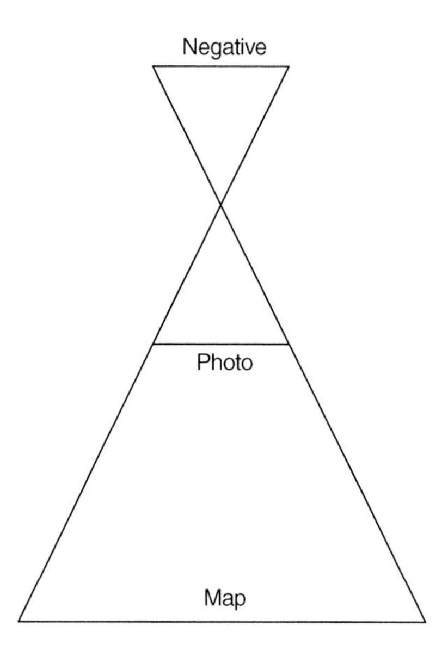

Figure 9.1 Map scale/photo scale relationship.

Calculation of photo scale is based upon horizontal accuracy objectives, and Equation 9.3 should be employed.

$$s_p = s_m * f_x \tag{9.3}$$

where:

s_p = photo scale denominator (feet)
s_m = map scale denominator (feet)
f_x = enlargement factor from photo to map scale

The relationship of photo scale and map scale is visually demonstrated in Figure 9.1.

9.2.3 Topographic Features

A given photo scale must maintain accuracy of the selected contour interval by reference to the C-factor. Although the credence of a C-factor is debated in some quarters, it is universally accepted within the American mapping community. Planners should be judicious in selecting a C-factor that will maintain map accuracy.

9.2.3.1 Flexible C-Factor

Currently in the United States, the flexible C-factor may diverge significantly from one production system to another (i.e., analytical stereoplotter vs. softcopy workstation). The selected factor may be subject to qualitative analysis and be

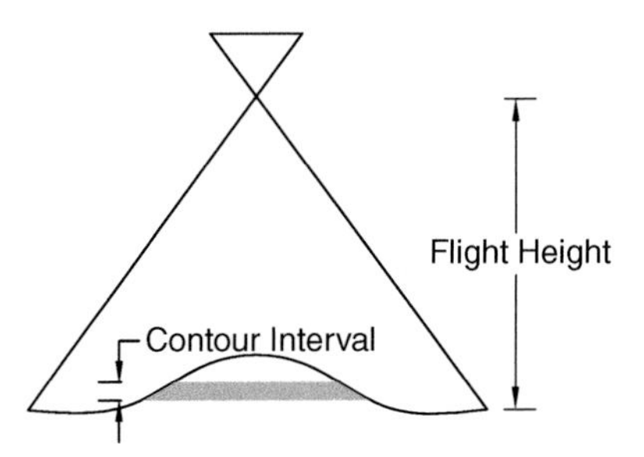

Figure 9.2 Relationship of the contour interval to flight height.

influenced by available equipment, compiler, or the philosophy of the organization providing the mapping product. It is better to be conservative in the application of the C-factor, since the production cost differential may be of less consequence than potential future liability which may be incurred from "stretching" the accuracy limits of the imagery.

The drawing in Figure 9.2 graphically illustrates the relationship between the contour interval and flight height. Once the flight height has been determined, the resultant negative scale can be reckoned by entering the appropriate variables into Equation 6.3 (Chapter 6).

9.2.3.2 Photo Scale/Contour Interval

Calculation of the photo scale based upon the vertical accuracy is premised on the interaction of the C-factor and contour interval.

Purpose of Mapping

The appropriate contour interval is dependent upon the purpose of the mapping. Refer to Table 9.1 to determine which contour interval meets the criteria for the topographic mapping.

Accuracy

The accuracy of the selected contour interval relies upon the C-factor, which is dependent upon the mapping class and production system. ASPRS map classes and interpolated NMAS accuracy standards limit C-factors to those found in Table 9.2 for analytical stereoplotters. The references to softcopy workstation C-factors in Table 9.2 are based upon the author's current experiences and discussions with various photogrammetric mapping firms in the United States. FGDC NSSDA C-factor recommendations mirror either ASPRS standards or NMAS.

Flight Height

Once the contour interval/C-factor combination has been selected, the height of the aircraft can be calculated by Equation 9.2.

Photo Scale

The proper photo scale to maintain vertical accuracy can be computed using Equation 6.3 (Chapter 6).

9.2.3.3 Contours from Existing Photos

If the user has access to existing photographs, the appropriate contour interval associated with those photos can be determined. With the aid of a map measurement between two solid features and the corresponding ground distance between those same image features, the photo scale can be determined. Then, based upon a suitable C-factor for the stereomapping system utilized, the smallest contour interval would be determined by Equation 9.4.

$$c_i = H/c_f \qquad (9.4)$$

where:
c_i = contour interval (feet)
H = flight height (feet) above mean ground level
c_f = selected C-factor

9.2.4 Photo Scale Selection

The desired map scale/contour interval may or may not be compatible. This is precisely the reason that photo scale must be computed for both the map scale and contour interval.

Large-scale maps are usually discussed in engineers' scale rather than representative fraction (1 in. = 50 ft vs. 1:600). This is due to the use of large-scale maps by professional engineers, who prefer working with engineers' scale. Planners, cartographers, geographers, and other GIS users generally prefer horizontal scales stated in representative fraction.

9.2.4.1 Compatible Parameters

To illustrate the selection of photo scale, assume the scenario requiring a map scale of 1 in. = 50 ft (1:600) with contours at 1-ft intervals compiled on an analytical stereoplotter.

Horizontal

To satisfy the horizontal (planimetric) accuracy prescribed by NMAS, the enlargement factor for employing an analytical stereoplotter is 7.5 times (see Table 9.3). Inserting the map scale and enlargement factor into Equation 9.3 yields:

$$s_p = s_m * f_x = 50 \times 7.5 = 375, \text{ or } 1 \text{ in.} = 375 \text{ ft}$$

Vertical

Table 9.2 indicates that the maximum C-factor for satisfying NMAS when using an analytical stereoplotter is 2100.

To satisfy the vertical (topographic) accuracy, the flight height above mean ground level would be derived by Equation 9.2.

$$H = c_f * c_i = 2100 \times 1 = 2100 \text{ ft}$$

Solving Chapter 6, Equation 6.3, the resultant negative scale would be:

$$s_p = H/f = 2100/6 = 350, \text{ or } 1 \text{ in.} = 350 \text{ ft}$$

Selected Scale

When selecting the photo scale, it should be the larger of the two calculated scales so that accuracy can be assured for both horizontal and vertical features. In this example the selected photo scale would be 1 in. = 350 ft, although any scale between 1 in. = 350 ft and 1 in. = 375 ft should maintain both horizontal and vertical accuracy.

The horizontal and vertical scales are compatible because they are near to each other in absolute value.

9.2.4.2 Incompatible Parameters

There are situations where the map scale and contour interval are not directly compatible. This situation of incompatibility does not preclude mapping. Rather, it forces the planner to choose the photo scale wisely.

Horizontal

Consider a project which specifies NMAS mapping to scale 1 in. = 200 ft (1:2400) with 2 ft contours on an analytical stereoplotter. To satisfy the horizontal (planimetric) accuracy prescribed by NMAS, the enlargement factor for employing an analytical stereoplotter is 7.5 times (see Table 9.3). Inserting map scale and enlargement factor into Equation 9.3 yields:

$$s_p = s_m * f_x = 200 \times 7.5 = 1500, \text{ or } 1 \text{ in.} = 1500 \text{ ft}$$

Vertical

Table 9.2 indicates that the maximum NMAS C-factor for an analytical stereoplotter is 2100. To satisfy the vertical (topographic) accuracy, the flight height above mean ground level would be derived with Equation 9.2.

$$H = c_f * c_i = 2100 \times 2 = 4200 \text{ ft}$$

Solving Equation 6.3 (Chapter 6), the resultant negative scale would be:

$$s_p = H/f = 4200/6 = 700, \text{ or } 1 \text{ in.} = 700 \text{ ft}$$

Considering these parameters, the choice of photo scale for compiling contours would be 1 in. = 700 ft, and the photo scale for compiling cultural features would be 1 in. = 1500 ft.

In this instance, the vertical aspect becomes of prime importance, since utilizing the smaller scale (1 in. = 1500 ft) photographs would not permit the collected digital topographic information to meet vertical accuracy requirements stipulated by NMAS.

In the same project suppose that the map scale is changed to 1 in. = 50 ft and all other parameters remain the same as above.

To satisfy the horizontal (planimetric) accuracy prescribed by NMAS, the enlargement factor for employing an analytical stereoplotter is 7.5 times (see Table 9.3). Inserting map scale and enlargement factor into Equation 9.3 yields:

$$s_p = s_m * f_x = 50 \times 7.5 = 375, \text{ or } 1 \text{ in.} = 375 \text{ ft}$$

Table 9.2 indicates that the maximum NMAS C-factor for an analytical stereoplotter is 2100. To satisfy the vertical (topographic) accuracy, the flight height above mean ground level would be derived with Equation 9.2.

$$H = c_f * c_i = 2100 \times 2 = 4200 \text{ ft}$$

Solving Equation 6.3 (Chapter 6), the resultant negative scale would be:

$$s_p = H/f = 4200/6 = 700, \text{ or } 1 \text{ in.} = 700 \text{ ft}$$

Using these parameters, the photo scale for compiling contours would be 1 in. = 700 ft, and the photo scale for compiling cultural features would be 1 in. = 375 ft.

In this instance, the horizontal aspect becomes of prime importance, since utilizing the smaller scale (1 in. = 375 ft) photographs would not permit the collected digital planimetric information to meet horizontal accuracy requirements stipulated by NMAS.

9.3 PLANNING AN AERIAL PHOTO MISSION

Now that the elementary geometric principles of aerial photography have been explored, this knowledge can be used in planning a photo mission. The intent is to plan a photo mission to cover the area delineated in Figure 11.4 (Chapter 11) with photos to scale 1 in. = 500 ft, assuming sidelap to be 30%, forward overlap to be 60%, and the camera focal length to be 6 in.

9.3.1 Laying Out Flight Lines

Initially, it is necessary to calculate the distance between flight strips.

9.3.1.1 Sidelap Gain

If the photos are to have a sidelap of 30%, the sidelap gain is calculated by Equation 6.7 (Chapter 6).

$$g_{side} = s_p * w * \left[\left(100 - \text{percent}_{side}\right)/100\right]$$

$$g_{side} = 500 \text{ ft} * 9 \text{ in.} * \left[\left(100 - 30\%\right)/100\right] = 3150 \text{ ft}$$

As an example of flight line layout, assume that the total width of the project is 10,500 ft as measured from an adequate map source. If strips are 3150 ft apart, this project will need 3.33 flight strips (10,500/3150 ft). It is impossible to have a fractional strip, so this project will require four photo lines. Flight lines are laid out leaving an equal distance along the east and west edges as noted in Figure 11.4 (Chapter 11).

9.3.1.2 Flight Line Orientation

In this situation, since the area is almost square, lines will be flown in the north/south direction to take advantage of the road system for accessing field control layout.

If possible, flight strips are usually oriented in a cardinal direction. However, this is not always practical because sometimes the project area may be skewed. Flight lines should be oriented parallel to the long dimension of the project if possible, since it will probably require fewer flight lines. It may be desirable to position flight lines relative to easiest access (roads, transmission lines, and cleared strips in woods) for field control routing. Ease of placement of field control should not be the overriding consideration for flight planning; however, field control can be a significant cost factor in a mapping project. Accessibility for ground survey crews can affect the time and cost associated with any mapping project. In states which employ the land subdivision system discussed in Chapter 10, flight lines may possibly be oriented so that the pilot can take advantage of sectionalized lines for guiding the flight course. However, this was of more consequence when flights were controlled visually, where pilots could make use of obvious land subdivision lines as reference guides. This is of less importance when ABGPS systems are utilized, since the aircraft can more easily maintain course through electronic piloting.

9.3.1.3 Airborne Global Positioning System (ABGPS) Navigation

ABGPS navigation of aerial photography missions is the current standard practice for all but very small projects. Software and hardware systems are available that fully automate the flight planning steps and processes and maximize the efficiency of aerial photo missions. Hardware and software requirements include CADD workstations, GPS antennas, and processing units. These systems require the mission planning team to develop the coordinates of the beginning and ending points for each flight line and to plot them on a digital reference map. The reference map is stored in an on-board computer navigation system tied to a GPS antenna and positioning system. The pilot uses the GPS to navigate the aircraft and activate the camera at the proper time and intervals. These systems allow for efficient mission planning. Utilizing these ABGPS navigation systems also provides a limited amount of quality control by requiring that the pilot use the mission planning data to guide the craft and the photography collection.

9.3.2 Determining Number of Photos

After the lines are laid out on the flight map as shown in Chapter 11, Figure 11.4, it is necessary to calculate the number of exposures required to provide stereophoto coverage.

If the photos are to have a forward overlap of 60%, the endlap gain is calculated by Equation 6.5 (Chapter 6).

$$g_{end} = s_p * w * \left[\left(100 - \text{percent}_{end}\right)/100\right]$$

$$g_{end} = 500 \text{ ft} * 9 \text{ in.} * \left[\left(100 - 60\%\right)/100\right] = 1800 \text{ ft}$$

To compute the number of photos on each line, divide the length of line by the net gain (17,000/1800), which equates to 9.4 photos per flight line. Since there are no fractional photos, each line will contain ten photos. At least one additional photo for each flight line is necessary to assure stereoscopic coverage on the total area. This means that the project will require a total of at least 44 exposures.

9.3.3 Calculating Flight Height

Finally, flight height above mean ground level of the aircraft to maintain the desired photo scale should be determined from Equation 6.4 (Chapter 6).

$$H = s_p * f = 500 \text{ ft} * 6 \text{ in.} = 3000 \text{ ft}$$

Referring to the project map in Figure 11.4 (Chapter 11) it appears that the terrain elevation varies between 550 ft and 650 ft for an average ground level of about 600 ft. Then:

$$\text{flying altitude} = \text{flight height} + \text{average ground}$$

$$= 3000 \text{ ft} + 600 \text{ ft} = 3600 \text{ ft above mean sea level}$$

Knowing the flying altitude and the location of individual flight lines, the crew can fly this project.

CHAPTER 10

Geographic Referencing

10.1 GEOGRAPHIC LOCATION SYSTEMS

To create a reliable map, it is necessary to relate the digital data to its true terrestrial situation.

10.1.1 Land Subdivision

The system of geographically parceling land in the United States has undergone changes since the American Revolution.

During the American colonial period, land subdivision was accomplished by metes-and-bounds surveys. Irregularly shaped parcels were segregated by surveying between visible features on the ground, such as boulders, piles of stone, trees, or fence corners. With the passage of time these objects had a way of disappearing, and retracing landlines has been difficult if not impossible.

Land in the United States, other than the 13 original colonies, is divided into units of the System of Rectangular Surveys.* Land is still parceled by the System of Rectangular Surveys on a variety of map sources in the United States. Anyone using county tax assessor's plat books, USGS quadrangle sheets, U.S. Soil Conservation soil type maps, U.S. Forest Service timber type maps, and other similar documents will come in contact with this land subdivision procedure.

Between 1803 and 1956 a number of initial reference points were established throughout the United States. In some localities a single reference point may serve more than one state. There are also situations where several initial reference points fall within a single state. For instance, Alaska contains five reference points. Using these points as initial surveying references, land is broken into townships, each 6 mi^2, which are further divided into 36 sections, each containing 1 mi^2 (640 acres). Figure 10.1 indicates the standard pattern in which sections within a township are numbered.

* For a more detailed narration of land subdivision refer to Chapter 10 in *Aerial Mapping: Methods and Applications,* Lewis Publishers, Boca Raton, FL, 1995.

Township Line

Range Line

6	5	4	3	2	1
7	8	9	10	11	12
18	17	16	15	14	13
19	20	21	22	23	24
30	29	28	27	26	25
31	32	33	34	35	36

Figure 10.1 Township and range identification on a USGS quadrangle sheet.

Figure 10.2 is a copy of a segment of a USGS quadrangle sheet showing the System of Rectangular Surveys. Note that the upper left corner R2E (along top margin) and T41N (along left margin) designate range and township coordinates, and the number group (36, 31, 1, 6) indicates the intersection of four individual sections.

When sections are subdivided into fractions for land parceling, the description begins at the smallest unit and works upward through the hierarchy. Contrarily, the creation of a unit works in reverse order. The 160-acre tract noted in Figure 10.3 would be identified as the NW/4 of Section 7, T6N, R11W.

10.1.2 Digital Mapping Data

A significant proportion of digital data generated by aerial surveys will eventually find their way into databases. Information derived from remote sensors is also incorporated into databases. Data from these and other sources can be mixed and matched in information system projects (Chapter 13). Hence, individual segments of the digital data must be keyed to the same geographic planes in order to be harmonious. Neither the metes-and-bounds nor the rectangular survey protocols are suitable for controlling contemporary digital mapping projects.

Digital information that represents a map resides in a database matrix composed of a multitude of individual points. Each point is spatial, having a coordinate triplet value for X (easting), Y (northing), and Z (elevation). Data are referenced horizontally to one of several grid coordinate projection systems. Elevation should be referenced to mean sea level.

10.1.3 Coordinate Systems

Coordinates can be referenced to any desired precise geographic grid system. The selected system is at the discretion of the user. This can be a standard reference

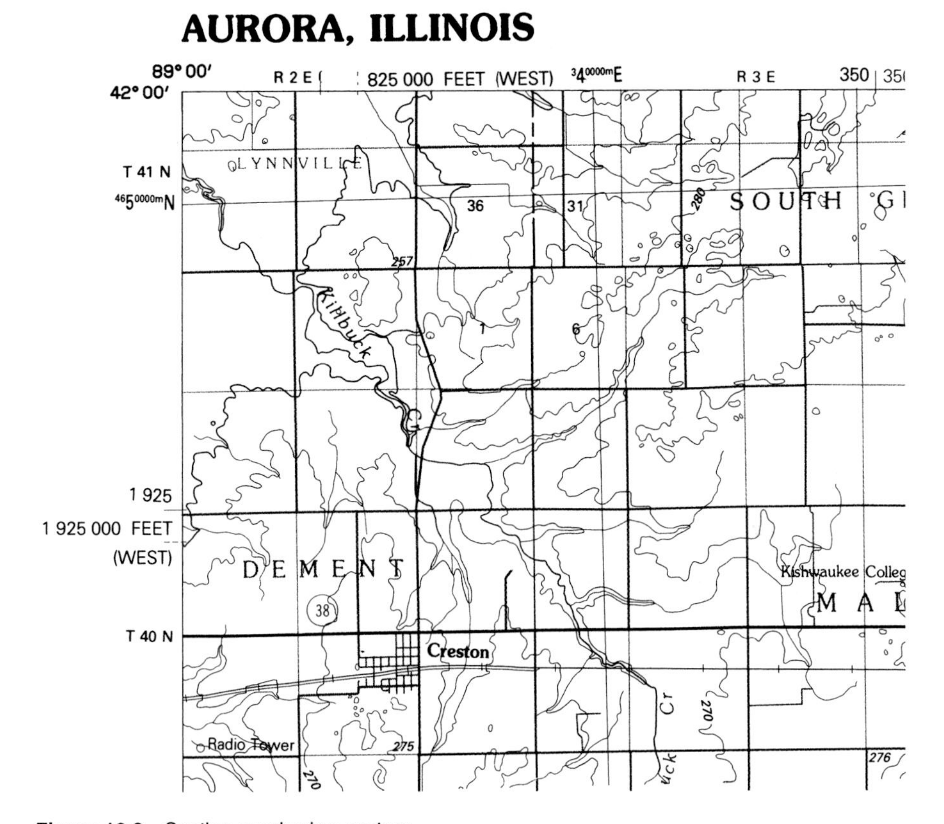

Figure 10.2 Section numbering system.

system to tie the mapping to the world or individual state, or it could be an assumed grid reference pertinent only to the specific site.

Data are increasingly incorporated into databases for information systems by diverse groups, and this trend will continue to grow in the future. If the data are to be inserted into a conglomerate of other packages of information, it must be translated to abide with the coordinate system of the whole.

It would be best to consider standard coordinate systems for collecting data to facilitate comprehensive use of the data. Three grid systems are popular for mapping within the United States, and all can be converted to match one another in a database. If one were to look at a quadrangle sheet, published by the USGS, he/she would note that all three of these coordinate systems are imprinted along its borders.

10.1.3.1 Universal Transverse Mercator

Universal Transverse Mercator (UTM), a metric (meters) grid structure, is a worldwide planar map projection which breaks the globe into 60 zones, each covering 6° of longitude. UTM is a cylindrical envelope intersecting the earth along two lines that are parallel to a central meridian. Data referenced to this grid system

ELSAH QUADRANGLE
ILLINOIS-MISSOURI
7.5-MINUTE SERIES (TOPOGRAPHIC)

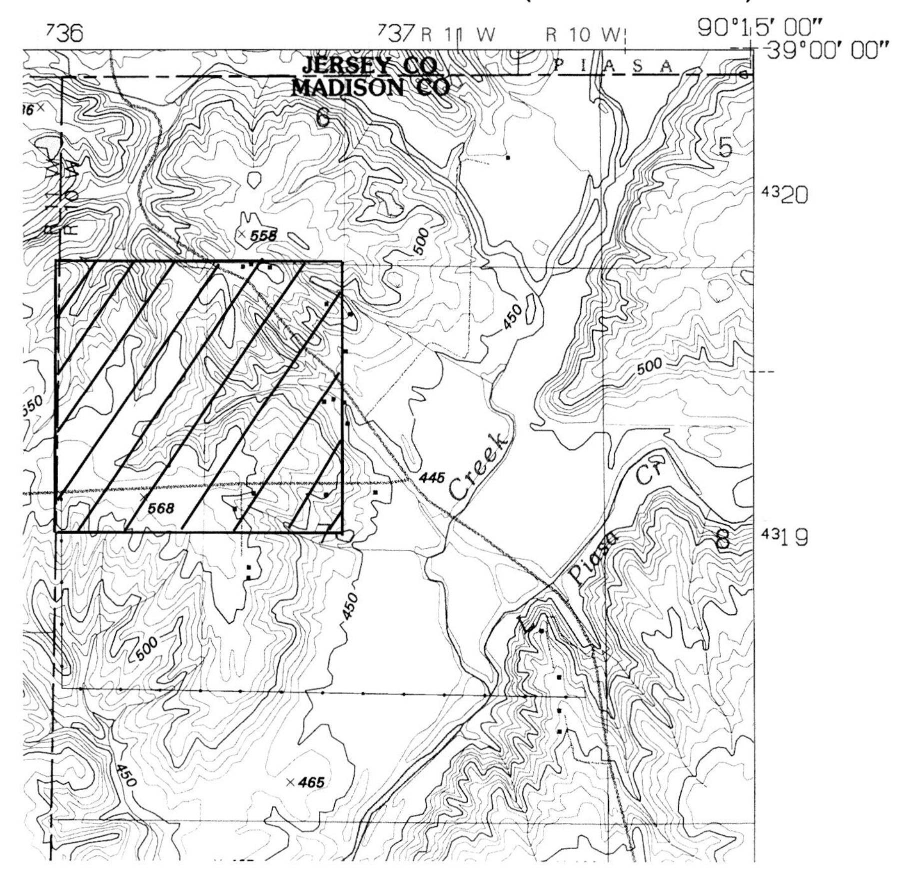

Figure 10.3 Land subdivision system tract identification.

can be correlated to worldwide mapping. Many GIS projects utilize this grid system to incorporate analogous data from diverse sources.

10.1.3.2 State Plane

The State Plane Coordinate System (SPCS), expressed in feet, segregates 120 zones throughout the United States. Conformal conic projection can be used in states that are wide in the east–west direction. Transverse mercator projection can be used in states that are narrow in the east–west direction. Much of the engineering mapping within state boundaries is controlled by this grid system.

10.1.3.3 Latitude/Longitude

Latitude/longitude coordinates are not normally used on large-scale mapping, but they are common to small-scale cartography and marine charting. If captured data are to be used in a national or international database, then this projection may be appropriate.

10.2 GROUND CONTROL SURVEYS

There are survey systems for referencing any tract of land to its true geographic placement on the earth.

10.2.1 Basic Surveying

Surveying is an ancient technology.* Mapping by either conventional or electronic methods requires the determination of spatial coordinates of specific strategically placed features on the ground. To arrive at these coordinates with conventional surveying techniques, three functions must be performed. Horizontal coordinates (X and Y) are derived by turning angles, and measuring distances and hypsometric measurements determine elevations (Z).

10.2.1.1 Angles

Horizontal angles must be sighted between successively selected points on the ground. For this procedure, magnetic compasses progressed into the transit — an instrument combining magnetic compass, horizontal azimuth card, and vertical angle vernier — then progress into the theodolite, a precision instrument for measuring horizontal and vertical angles.

10.2.1.2 Distances

In conjunction with turning angles, distances must be measured. The Gunther chain, which was probably preceded in some point in history by pacing or knotted string, developed into steel tapes for measuring distances. The age of electronics ushered in the electronic distance-measuring device. This apparatus emits light beams, at first visible and then laser, and measures the time it takes the beam to bounce off a reflector set over a distance point. A laser is a phased monochromatic beam of light that has been electronically stimulated.

10.2.1.3 Levels

Over the years elevations have been derived through the use of hand levels, dumpy levels, aneroid barometers, trigonometric levels, and differential spirit levels.

* A brief history of the technology of surveying is discussed in Chapter 11 in *Aerial Mapping: Methods and Applications,* Lewis Publishers, Boca Raton, FL, 1995.

10.2.2 Electronic Surveying

Electronic tacheometers (ETI), also termed total stations, incorporate both theodolite and electronic distance-measuring capabilities within a single instrument. Total stations are capable of calculating and outputting:

- Horizontal angle
- Vertical angle
- Horizontal distance
- Slope distance
- Vertical distance

When cabled to an electronic field book, total stations collect survey information directly in digital form. These data can be downloaded into a computer to perform coordinate geometry functions and topographic mapping.

10.3 GROUND SURVEY TOOLBOX

Locating features on the earth that can be seen and measured in the imagery is required in the photogrammetric mapping process. The required accuracy of the established feature locations is dependent upon the required map accuracy. Generally, the accuracy of the required feature locations must be more accurate than the mapping requirements for similar features that will be compiled in the spatial data collection. Therefore, the required map scale is an important consideration when deciding methods to be employed for ground control.

A good source for current ground survey accuracies and recommended standards and tolerances for various types of spatial data collection is the FGDC Geospatial Positioning Accuracy Standards, Part 4: Standards for Architecture, Engineering, Construction (A/E/C) and Facility Management, July 1998. Other considerations may include the following:

- Terrain within the mapping area
- Accessibility within and around the mapping area
- Vegetation and tree canopy within and around the mapping area
- Building heights and density within and around the mapping area
- Time and funding available to collect the ground control information

Therefore, the method employed to collect the data should not be based solely on "what is the newest technology." Surveyors have many types of equipment and methods available. Conventional traversing and level loops, as well as GPS methods, may be used. The decision should be based upon what survey methods within the surveyors toolbox fit the project. This chapter will not address detailed survey procedures and practices, but will discuss general requirements and information needed to make necessary decisions regarding types and methods to employ for a specific mapping project. A good source of current detailed information regarding conventional ground survey and GPS survey practices is the U.S. Army Corps of

Engineers, Engineering Manuals EM 1110-1-1002, Survey Markers and Monumentation and EM 1110-1-1003, NAVSTAR GPS Surveying.

10.3.1 Conventional Ground Survey

Many relatively small photomapping projects still rely upon conventional ground survey methods which include the use of survey transits, levels, and tapes to establish distances, horizontal locations, and elevations. These methods can produce location data to the accuracy required for most photogrammetry projects. The size, location, and terrain within the project and the time allotted for ground survey collection affect the staffing requirements when employing these methods.

10.4 GLOBAL POSITIONING

GPS methodology is often employed for establishing survey data. A GPS brings into play a surveying system that can isolate the position of a point on the earth's surface by making simultaneous observations on several orbiting NAVSTAR navigational satellites. Essentially, an electronic receiver measures the distances between the ground point and a minimum of four satellites, and the intersection of the divergent rays establishes the spatial coordinates of the observing station.

The courses of the two dozen operating satellites are predicted and tracked by the National Geodetic Survey. The anticipated ephemerides (positional) information is broadcast by the satellite. Several continuous tracking stations are scattered throughout the world, meticulously charting the paths of the satellites. Both the tracking data and broadcast information are available to the user, with the former providing more accurate data for processing receiver information.

10.4.1 Determining Spatial Coordinates

Determination of the coordinates of a ground station by GPS procedures relies upon intersection geometry. The GPS receiver, a pseudo-range measuring device, accepts carrier signals (microwaves with a set velocity) from multiple satellites, measures transmission time lapse, and determines distance from each satellite to the receiver. By measuring the carrier waves from a ground station to several satellite positions simultaneously, the XYZ coordinate can be determined. For ascertaining three-dimensional coordinates, the receiver must maintain a continuous lock on a minimum of four orbiting platforms simultaneously. Figure 10.4 depicts a point position at the intersection of four satellite ranges. R1 through R4 are measured pseudo-ranges between the receiver and satellites.

10.4.2 Global Positioning System Procedures

The accuracy requirement (horizontal and vertical) for the final mapping data sets establishes the ground survey accuracy requirements and thus the GPS methods

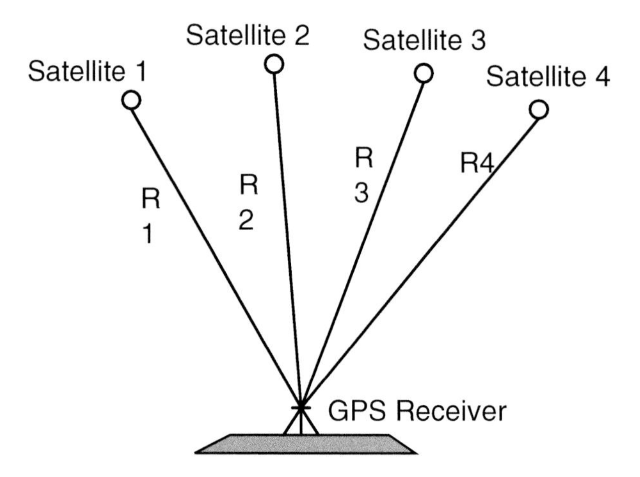

Figure 10.4 Intersecting satellite ranges.

that may be employed. Generally, for photogrammetric mapping, kinematic GPS methods will produce the feature location accuracy required. Some projects may require the establishment of selected features to an accuracy level only obtainable by static GPS methods.

When GPS is to be employed, the location of ground features for photo control must be carefully planned. A good reference for GPS planning is the U.S. Army Corps of Engineers, Engineering Manual EM1100-1-1003, NAVSTAR Global Positioning System Surveying.

10.4.2.1 Static Global Positioning System

In static traverses, at least two receivers must be used; both must be locked on to the same group of satellites. One receiver resides over a location with known coordinates, which could be a previous point in the circuit. The other receiver is set over a point with unknown coordinates. Observation time spent at each baseline point pair may span a significant period of time. After the observation is complete at the baseline pair, both receivers can be moved to new positions so long as one occupies a point with known coordinates. In this fashion, observations are made at all of the stations on the traverse in turn.

10.4.2.2 Kinematic Global Positioning System

As in static mode, kinematic traverses demand the inclusion of at least two receivers. One receiver must remain fixed over the same point, with known coordinates, during the entire survey period. The other receiver roves to each point of unknown coordinates in turn. Time spent by the roving receiver at each circuit point is much less than that used in static mode, perhaps as little as a few minutes.

10.4.3 Airborne Global Positioning (ABGPS)

Today many photogrammetric mapping projects employ ABGPS technology to minimize ground control collection for the mapping project. By locking on to several navigation satellites this device maintains a constant spatial positioning record of the sensing systems. Relating the film coordinates to a map allows the pilot to pinpoint the location of the aircraft on a specific exposure frame. This coupling of remote sensing and spatial positioning offers unique mapping capabilities.

ABGPS technology has been developed to a point where standard procedures and repeatable results make it almost standard practice for all medium to large project areas. GPS hardware and software vendors have teamed up with mapping hardware and software vendors to develop planning and processing tools that minimize errors and drastically reduce the time from aerial photography to a map file in hand.

10.4.3.1 Aircraft

Many private and public organizations currently mount GPS receivers in the airplane during a photo flight mission. This system allows the spatial fixing of the camera at the instant of exposure. Relating this position to the mapping site, it is possible to produce planimetric and/or topographic maps without the necessity of putting surveyors onto the site.

10.4.3.2 Reference to Ground Station

This system allows the ABGPS receiver to be interfaced with a camera system. In this procedure, the GPS in the aircraft is correlated with a static GPS ground station so as to relate the onboard receiver with the known ground station.

10.4.3.3 Aircrew Duties

The traditional concept of the duties of the aircrew members is changing. An aerial mission now can be considerably automated. Herein is described the general working of one such system. In an electronic guidance situation, both the pilot and the photographer must be conversant with computers and GPS.

Flight Map

The exterior boundary of the mapping site is digitized relative to an appropriate ground coordinate system. Flight parameters (photo scale, endlap, and sidelap) are input into the computer along with the digitized site boundary. Upon command, the computer will produce a hardcopy map delineating the location of each flight line upon the site map.

Preflight

Before takeoff, the flight map information is entered into an airborne personal computer which is linked with the onboard geographic positioning and camera systems. The flight map is displayed on the graphic screen in full view of the pilot.

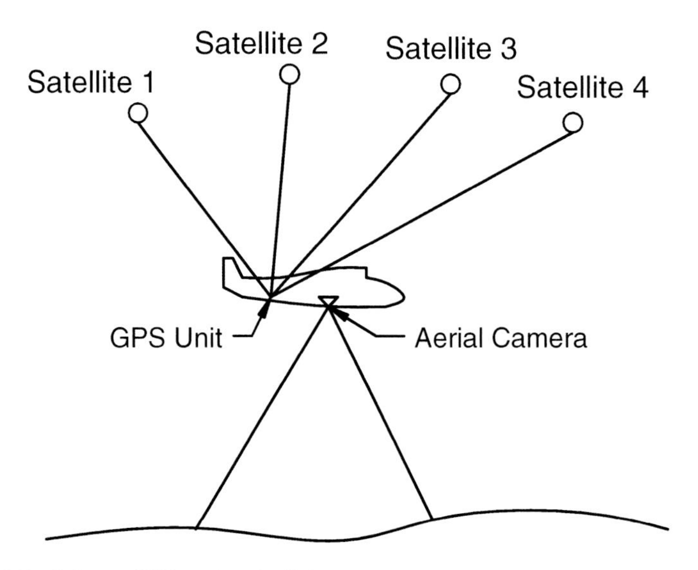

Figure 10.5 Airborne GPS system in flight.

Photo Mission

After takeoff, the pilot flies to the starting point of the initial flight line. Figure 10.5 is a schematic diagram of an ABGPS. As the aircraft proceeds along the intended flight path, its position is graphically superimposed on the screen. By observing the true position of the aircraft and the intended course, the pilot can adjust the heading of the aircraft to adhere to the designated flight path. At periodic intervals the camera automatically makes an exposure, and the location of the perspective center is displayed on the graphics monitor.

By relating the precise instant of mid-exposure recorded by the camera, the geodetic XYZ position of the camera perspective center can be determined from the data captured through interfacing with the GPS receiver.

The ABGPS, like other photogrammetric surveying and mapping tools, has unique limitations and requirements. The cost and experience of a photogrammetric mapping team are always considerations. The experience of the team and additional equipment does affect overall cost, quality, and time to complete a project.

Considerations

Some unique factors which influence the cost of photo missions should be considered when pondering the utilization of ABGPS:

- In the past, the aerial photo mission has been typically the least costly phase of a mapping project. With the application of ABGPS the cost of the aerial photographic phase will increase, perhaps significantly.
- Crew members must be proficient in computer manipulation and GPS procedures. This could require higher salaries.
- Airborne computers and GPS receivers present additional equipment costs.

10.5 BASIC CONTROL NETWORKS*

Prior to collecting ground survey data at photo control points, it may be necessary to accomplish basic survey circuits in order to reference them to existing networks.

10.5.1 Conventional Surveys

Initially, project control is referenced to an existing station for which appropriate coordinates and/or elevations have been established by the USGS or the NGS (National Geodetic Surveys). This may require the surveyor to run horizontal traverses or vertical circuits from points outside the mapping area. Field surveys for establishing horizontal control point information require that vector traverses be run through a series of basic control points. Field surveys for establishing basic vertical control point information require that a differential level circuit be run between successive stations.

Field surveys are links between aerial photography and map compilation; all are of equal relevance to project quality assurance. Any weak link in this "chain" can negate the judicious effort of the other links. The practice of establishing an unproven coordinate position or elevation of control points is detrimental to the scheme of quality control.

Once the basic control has been established, the surveyor then runs spur traverses or circuits from the temporary turning points or benchmarks to individual photo control wing points.

10.5.2 Control Reference

For aerial mapping to be functional for the map/data user, the surveyor must reference the control data to legitimate established geographic data.

10.5.2.1 Horizontal

Coordinates are derived from the azimuth and distance of each tangent as measured in the field for whatever grid system to which the mapping is to be referenced. Computed coordinates can refer to whatever grid system the user specifies, but there are two standard systems which are probably most commonly employed, State Plane Coordinate System (SPCS) and Universal Transverse Mercator (UTM).

State Plane Coordinate System

Each state has at least one grid zero reference point, and state plane coordinates are read as northings and eastings, in United States feet, from the relevant reference.

Universal Transverse Mercator

The UTM locates northing and easting, in metric units, coordinates within various designated global zones.

* An expanded view of basic control networks can be found in Chapter 11 in *Aerial Mapping: Methods and Applications,* Lewis Publishers, Boca Raton, FL, 1995.

10.5.2.2 Vertical Control

Most mapping projects require that elevations be referenced to a known vertical datum by commencing and closing on established benchmarks, which may be a considerable distance from the mapping site.

10.5.3 Traverse/Circuit Accuracy

The Federal Geodetic and Control Committee (FGCC) has established accuracy limitations for first-, second-, and third-order horizontal and vertical surveys. These standards are used in surveys requiring sophisticated geodetic control.

10.5.3.1 Horizontal

For the largest scales, third-order, class I accuracy should suffice for horizontal control efforts on most mapping projects; for smaller scales, third-order, class II accuracy should be sufficient. FGCC horizontal accuracy standards stipulate accuracies noted in Table 10.1. Equation 10.1 defines distance accuracy in terms of a ratio in the form of 1:a.

$$a = d/e_{xy} \tag{10.1}$$

where:

a = distance accuracy
d = horizontal distance between survey points
e_{xy} = standard error of horizontal point pair

Table 10.1 FGCC Horizontal Accuracy Standards

Survey Class	Minimum Distance Accuracy
First order	1:100,000
Second order, class I	1:50,000
Third order, class I	1:10,000
Third order, class II	1:5,000

Table 10.2 FGCC Vertical Accuracy Standards

Survey Class	Maximum Elevation Difference ($mm/\sqrt{km}$)
First order, class I	0.5
First order, class II	0.7
Second order, class I	1.0
Second order, class II	1.3
Third order	2.0

10.5.3.2 Vertical

Third-order, for larger contour intervals, or second-order (class II), for smaller contour intervals, accuracy should suffice for vertical control efforts on most mapping projects. FGCC vertical accuracy standards stipulate accuracies noted in Table 10.2. Equation 10.2 defines elevation difference accuracy.

$$b = e_z / \sqrt{d} \tag{10.2}$$

where:
- b = elevation difference accuracy ratio
- d = distance between survey points (kilometers)
- e_z = standard error of difference between vertical points (millimeters)

10.6 PHOTO CONTROL POINTS

Prior to commencing mapping from aerial photos, ground survey information is required on specific terrain features in order to relate the photogrammetric spatial model to its true geographical location. These terrain features may be portrayed in two ways: by identifiable photo image features or ground targets (panels). Some project conditions may dictate that a combination of image points and targets can be most realistically employed.

Acquisition of ground control data on photo image points is a necessary requirement for photogrammetric mapping for two primary reasons:

- To georeference the base imagery prior to spatial data collection
- To check the accuracy of the spatial data collected

Technology is constantly changing and adding to the tools that can be used to collect ground control. Recent advances in GPS are the underlying reason for many of the advances in ground survey methods, as well as photogrammetry in general.

10.6.1 Photo Image Points

Photo image points must be readily identifiable pictorial image features that are selected after the aerial flight is completed or pretargeted prior to the photo mission.

10.6.2 Ground Targets

A target is some kind of a panel point that is placed on unobstructed ground prior to photography. Figure 10.6 illustrates the effective presentation of a ground target (X) on an aerial photo. Ground targets create a discrete image point and can perhaps lead to better map accuracy. Targets stand less chance of being misidentified than an image feature, either by the surveyor or stereocompiler.

Figure 10.6 Ground target on an aerial photograph. (Courtesy of U.S. Army Corps of Engineers, St. Louis District.)

10.6.3 Size

A target must be of sufficient size to be recognized on the image. Dimensions of a target in the shape of a cross are easily computed.

Equation 10.3 defines the width of the legs of a typical ground target, as diagrammed in Figure 10.7, for any photo scale.

$$w = s_p * 0.002 \tag{10.3}$$

where:

w = width of target legs (feet)
s_p = scale denominator (feet)

Equation 10.4 defines the length of the target legs of a typical ground target, as diagrammed in Figure 10.5, for any photo scale.

$$l = 10 * w \tag{10.4}$$

where:

l = length of each cross arm (feet)
w = width of target legs (feet)

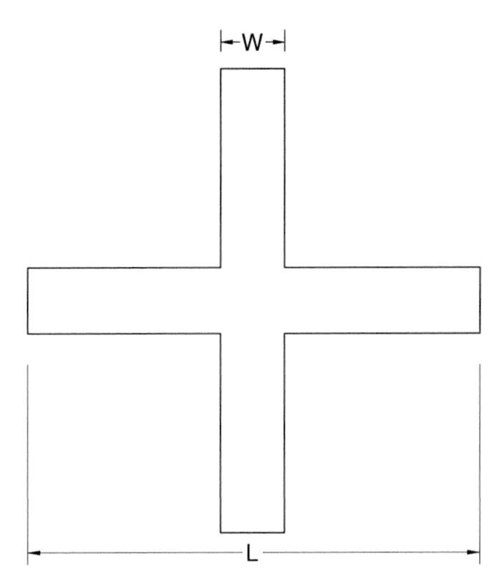

Figure 10.7 Typical ground target.

As an example, assume that a project requires a photo scale of 1 in. = 500 ft, the target size would be:

$$w = s_p * 0.002 = 500 \text{ ft} * 0.002 = 1 \text{ ft wide}$$

$$l = 10 * w = 10 * 1 \text{ ft} = 10 \text{ ft long, each cross arm}$$

10.6.4 Control Point Selection

To produce mapping from stereomodels, the aerial photo image must be scaled and leveled so as to cause the model to be georeferenced to a true geographic ground location. To do this, it is necessary to accurately relate the photos to the ground, both horizontally and vertically. This is done by establishing coordinates on specific horizontal "scale" points and elevations on vertical "level" points at the mapping site. In current mapping procedures, most points contain both horizontal and vertical information and are used for both scaling and leveling. The density of photo control points must be no less than four on each stereomodel, and the pattern should be one point in each of the corners, similar to the pattern illustrated in Figure 10.8.

10.6.4.1 Conventional Control

Conventional photo control point patterns require that spatial (XYZ) coordinates for every photo control point must be gathered on the ground. In conventional control point surveys, the photographs are exposed first, and then each required field control point is selected as an identifiable photo image point. The surveyor locates these points on the ground and gathers the appropriate field information.

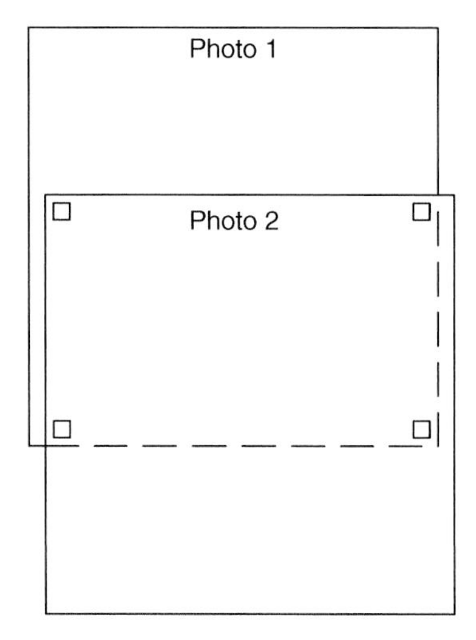

Figure 10.8 Photo control point requirements per stereomodel.

10.6.4.2 Skeletal Control

Skeletal surveys are run when the intent is to use aerotriangulation procedures to generate mapping photo control. In this situation, a lesser number of field control points are required than that needed in a conventional control pattern. Figure 10.9 illustrates a mapping project involving three successive stereomodels. Figure 10.10 notes the amount of conventional control that would be required to map the site.

By comparison, Figure 10.11 suggests the amount of skeletal field control needed to resolve the supplementary photo control bridging. The process to accomplish this will be outlined in Chapter 11 ("Aerotriangulation").

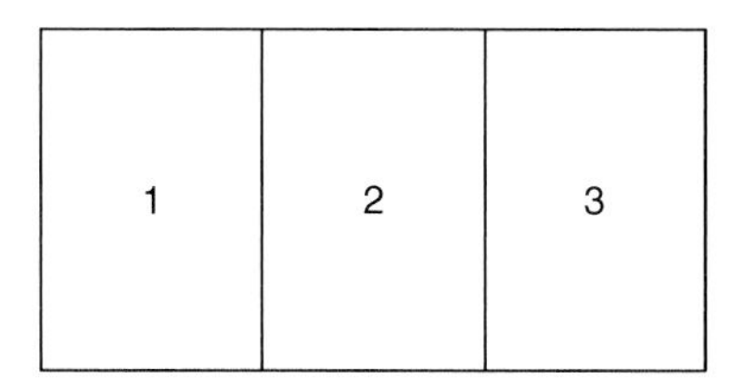

Figure 10.9 Three-model mapping site.

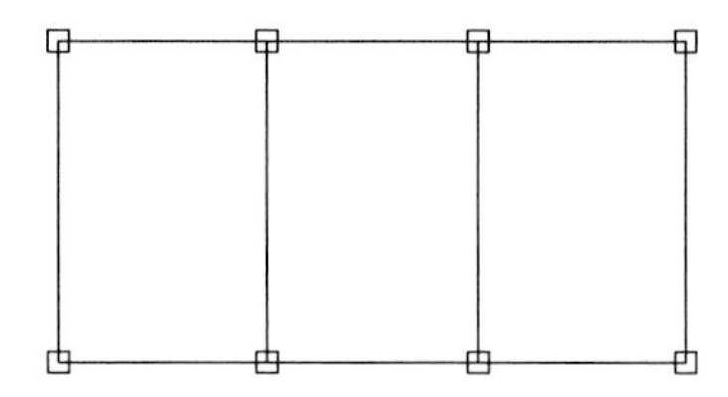

Figure 10.10 Conventional control point pattern on a triple model strip.

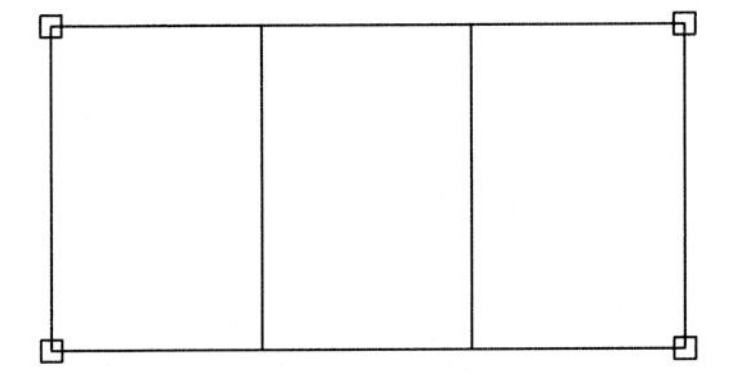

Figure 10.11 Skeletal control point pattern necessary to employ aerotriangulation on a three-model strip.

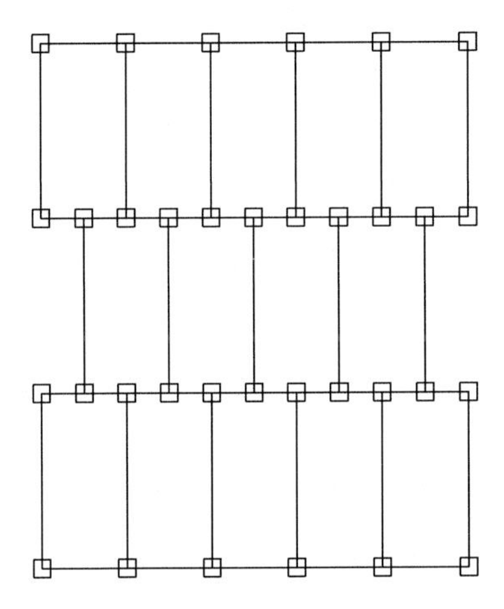

Figure 10.12 Conventional photo control point pattern on a multiple strip project.

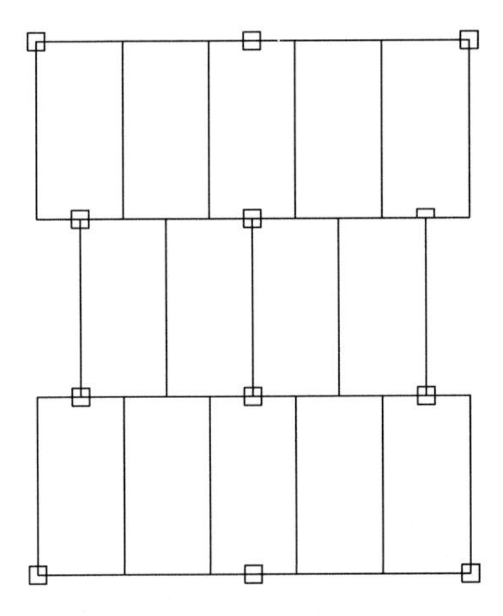

Figure 10.13 Skeletal photo control point pattern on a multiple strip project.

Broadening the scope of the mapping project to include several stereomodels on three adjacent flight lines, Figure 10.12 represents the amount of obligatory conventional control, while Figure 10.13 depicts the skeletal pattern which would satisfy the same qualifying factors.

CHAPTER 11

Aerotriangulation

11.1 PHOTO CONTROL BRIDGING

A critical phase in photogrammetric mapping is rectifying the aerial images to the appropriate tract on the surface of the earth. This is accomplished by collecting horizontal and vertical data, as discussed in Chapter 10, to ascertain the spatial location of a number of features that are visible and measurable on the aerial images. The process is often called control bridging, which refers to passing horizontal and vertical information from one aerial image to the next.

Geometric stability requires that a minimum of four points with established horizontal and vertical location spaced in the corners of a full stereomodel be used to fully rectify the model as shown in Chapter 10, Figure 10.6. On a project involving a few stereomodels this may be a conventional ground surveying enterprise, but it can be a costly ground survey venture for most medium to large project areas. The expense and time required to collect the ground survey data in this manner may render the mapping project impractical.

Computer processing has played a major role in driving mapping scientists to develop rigorous and efficient mathematical protocols that allow for the densification of stereomodel control from a minimal number of strategically positioned ground survey points. This procedure is generally referred to as aerotriangulation. Analytical software available today, with its built-in quality checks, has made aerotriangulation the preferred method of image adjustment to the earth for photogrammetric mapping. This chapter will not discuss the theory of these processes, but rather it will provide necessary guidance and explanation of procedures to plan and estimate the efforts required to perform satisfactory aerotriangulation for a photogrammetric mapping project.

11.1.1 Control Point Selection

To georeference a spatial stereomodel to the ground, it is necessary to furnish at least a minimum amount of both horizontal and vertical field control points.

To illustrate a conventional control solution, Figure 10.8 (Chapter 10) denotes a mapping project covering three photo flight lines and the pattern of survey points required to fulfill the minimum of four points per stereomodel necessary to assure a reliable spatial solution. Due to terrain conditions, access restrictions, or lack of identifiable cultural features, it is not always feasible to conform to these control point patterns.

Early on there was analog methodology to accomplish photo control bridging, but the advent of computer technology changed photo control extension procedures forever. Software vendors have created protocols that allow for automation of data input, error checking, and easy-to-read data output. Software has also been developed to handle the peculiarities of ABGPS control.

An aerotriangulation solution simultaneously determines the spatial intersection of image rays of a finite point from its position on overlapping photos. It is the analytical procedure that allows a mapper to utilize a skeletal pattern of field survey control to analytically generate sufficient photo image points to map a project.

11.1.2 Bridging Spans

Aerotriangulation allows many models to contain no field survey information. In practice, there are certain maximum uncontrolled model spans. They should be limited to those noted in Table 11.1. Figure 11.1 presents a comparison of the relative

Table 11.1 Maximum Uncontrolled Models for Aerotriangulation

Map Accuracy	Uncontrolled Models	
	Vertical	Horizontal
ASPRS1	2.0	4.0
NMAS	2.5	4.5
ASPRS2	3.0	5.0
ASPRS3	3.0	6.0

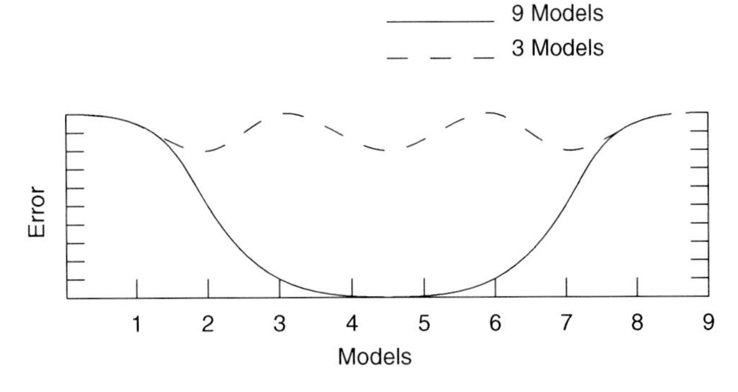

Figure 11.1 A comparison of the relative vertical errors propagated on a flight line spanning nine stereopairs.

vertical errors propagated on a flight line spanning nine stereopairs. The solid line represents a flight line spanning nine models, all of which are uncontrolled except for field control points at the extremities of the two end models. Compare this curve with the dashed line depicting the greatly reduced relative magnitude of errors if field control points were positioned at three-model intervals.

The acceptable feature types for field control points to be used in an aerotriangulation adjustment are the same as those described for conventional control. However, the number of field control points employed is substantially less and the spacing between them can be significantly larger. This procedure, when employed properly, should reduce field survey time and costs. To ensure this the aerotriangulation procedure involves a cost factor, but user-friendly software and high-speed computer processors have significantly lowered the time/cost element required to analytically adjust stereoimages to the earth. The cost of using aerotriangulation over less rigorous methods for most projects is also not a large percentage of the total cost of a mapping project today. Most major mapping contractors prefer to use this method because of the ease of use and the quality checks that the software can provide.

11.1.3 Skeletal Ground Control

Refer again to the conventional ground control point density in Chapter 10, Figure 10.10. The amount of skeletal horizontal ground control required to meet FGDC NSSDA, ASPRS, and NMAS skeletal field control requisites on the same site is diagrammed in Chapter 10, Figure 10.11. There are situations where breaking a large project into more than a single analytical run is advisable. In this event, ground points must be selected for each of the sectors as a separate entity.

It should be noted that photo image points, which fall outside the field control pattern, are subject to unusual errors. In this situation, the analytical bridge is actually extrapolating a solution whose accuracy tends to deteriorate rapidly. The further outside the control pattern the point falls, the greater the error. Therefore, it is desirable to have exterior photo control points located outside the mapping area.

11.1.4 Photo Control Extension Procedure

Photo control extension relies on utilizing differential parallax to analytically create coordinate sets on a number of selected photo image points. This process is designed to generate a sufficient pattern of photo image points based upon data collected on a minimal number of field survey stations. Several techniques accomplish photo control bridging. Analog procedures require a significant amount of manual manipulating in several stages, while bridging with softcopy instruments requires relatively little manual intrusion. So generally, the more manual bridging operation proceeds similarly to the procedure described herein.

11.1.4.1 Photo Image Point Location

A pattern of at least six photo image point locations per stereomodel, which equates to a minimum of nine points on most photos, is selected. Figure 11.2 suggests

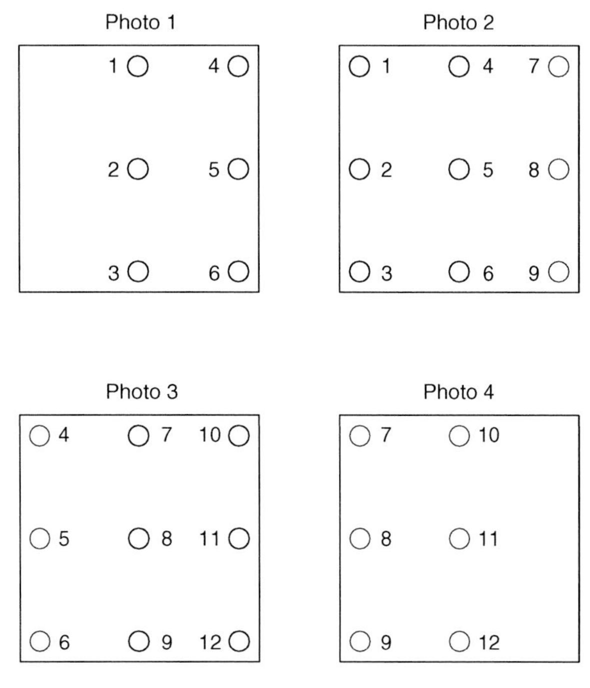

Figure 11.2 Basic patterns of at least six photo image point locations per stereomodel, which equates to a minimum of nine points on most photos.

the basic selection patterns of these arbitrary locations. Other pertinent points, such as field survey control points and strip connector pass points between adjacent flight lines, are also included.

11.1.4.2 Point Pugging

In the analog process an artificial target, called a pug point, is drilled into the emulsion of the diapositive at the location of each control point that appears on the image by employing a transfer device, as shown in Figure 11.3.

When utilizing a softcopy instrument, the control points are arbitrarily selected on the image which is displayed on the screen. Since the computer keeps these locations in its memory, there is no need for pugging.

11.1.4.3 Reading Diapositive Coordinates

Several pieces of information must be read from the pugged diapositives, utilizing a comparator device, into the aerotriangulation software prior to the processing. Once the point locations are selected on the image in the softcopy process, the computer keeps these locations in its memory, so there is no need for reading coordinates.

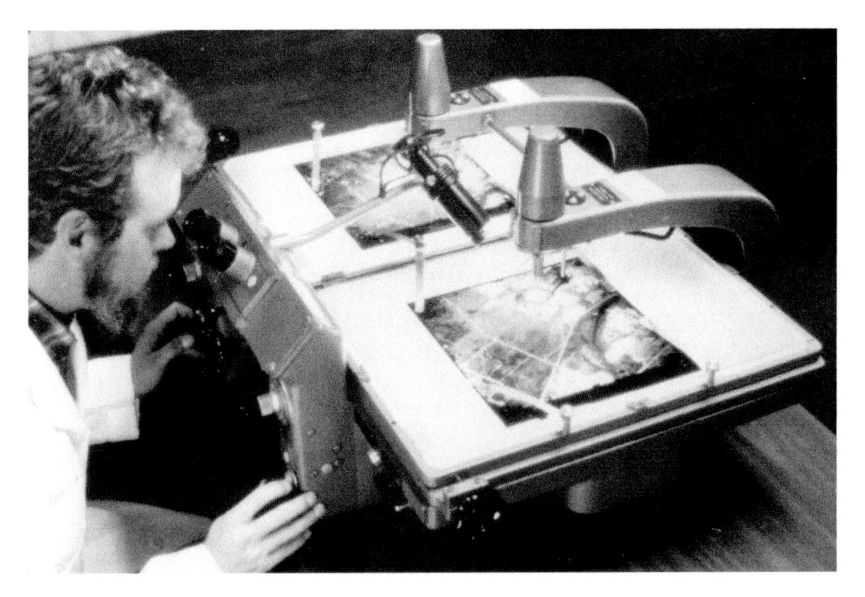

Figure 11.3 Point transfer device in use. (Courtesy of Surdex Corporation, Chesterfield, MO.)

11.1.4.4 Computer Processing

These raw plate coordinates, along with the field control point information, are imported into the computer and processed through an aerotriangulation package, of which there are a number available. A general procedure would include:

- In the sequential model assembly each stereomodel is processed through a relative orientation routine involving a least squares adjustment of colinearity equations. This solution produces individual model coordinates unrelated to any reference outside the matrix. Residuals of the model coordinates provide a test of the proficiency of the pugging and comparator readings.
- A strip formation procedure joins the independent stereomodels through a three-dimensional transformation. A series of equations links successive models by common pass points. The coordinates, at this stage, remain in an arbitrary reference scheme.
- Each strip undergoes a polynomial adjustment which produces preliminary ground coordinates for all of the photo control points.
- A simultaneous bundle adjustment provides a fully analytical aerotriangulation solution (as required for NMAS, ASPRS1, and ASPRS2), and the entire block of data passes through an iterative weighted least squares adjustment until a convergent "best fit" solution is attained. An RMSE error is noted so that the observer can judge how far the coordinates of each point were mathematically "stretched" out of position in order to resolve a solution.

11.1.5 Accuracy of Aerotriangulation

ASPRS map accuracy standards declare that the maximum horizontal and vertical RMSE of all the field control points should fall within that calculated by

Table 11.2 Accuracy Criteria for Aerotriangulation

Map Accuracy	Horizontal	Vertical
ASPRS1	10,000	9,000
NMAS	9,000	7,500
ASPRS2	8,000	6,000
ASPRS3	6,000	4,500

Equation 11.1. This equation can be used to calculate errors of X and Y coordinates and/or elevation.

$$E_{rms} = H/f_{xyz} \tag{11.1}$$

where:
E_{rms} = RMSE
H = flight height
f_{xyz} = factor specified in Table 11.2

There is a further, notable element to be considered. Since an RMSE may be roughly equated with an absolute average, be aware that there are errors at individual field points that will exceed the magnitude of the RMSE. Maximum error at any field control point, either horizontal or vertical, should not exceed triple the magnitude of the RMSE.

To pass judgment as to the validity of the RMSE, select the appropriate accuracy factor (dependent upon the desired accuracy goal) from Table 11.2. By entering this factor into Equation 11.1, the relevant maximum RMSE will become apparent.

11.1.6 Accuracy Check

There are a few items contained in the bridging printout that should be valuable in analyzing the relevance of the aerotriangulation:

- If the RMSE of the discrepancies at the field control points is significantly greater than that produced by Equation 11.1, there may be areas of intolerable inaccuracies.
- If the magnitude of the residual error at individual field control points is greater than triple the RMSE, there may be areas of intolerable inaccuracies.
- In the event that several field control points or points in critical locations are excluded from the final analytical solution, in order to gain a tolerable RMSE, there may be an adverse field control solution.
- If the analytical solution is the product of bare minimum control, the residuals may mask a faulty resolution.

11.1.7 Effects of Analytical Error

As noted by the fact that aerotriangulation networks generate an RMSE error, this analytical bridging procedure will introduce more error into individual photo control points than if data for these same points had been established by field surveys. This could have an effect on choosing a photo scale.

11.1.7.1 C-Factor Adjustment

It is advisable to consider adjusting the C-factor, by utilizing Equation 11.2, on projects that employ analytical control bridging procedures.

$$c_{fx} = c_f * \left[1 - \left(0.05 * m_{uc}\right)\right] \tag{11.2}$$

where:

c_{fx} = revised C-factor
c_f = original C-factor
m_{uc} = number of uncontrolled models spanned

11.1.7.2 Ground Targets

It would be good practice to consider ground targets for field control points. The use of preflight ground targets provides better definition to field control points and could help reduce aerotriangulation network residuals.

11.1.8 Example of Field Control Point Scheme for Aerotriangulation

Figure 11.4 shows a typical ground control plan. The plan shows the project boundary and flight lines as well as the approximate location or feature to be used as a ground control point most often established in the field by a surveyor. Spacing between control point locations in this example is approximately 2.5–3.0 models.

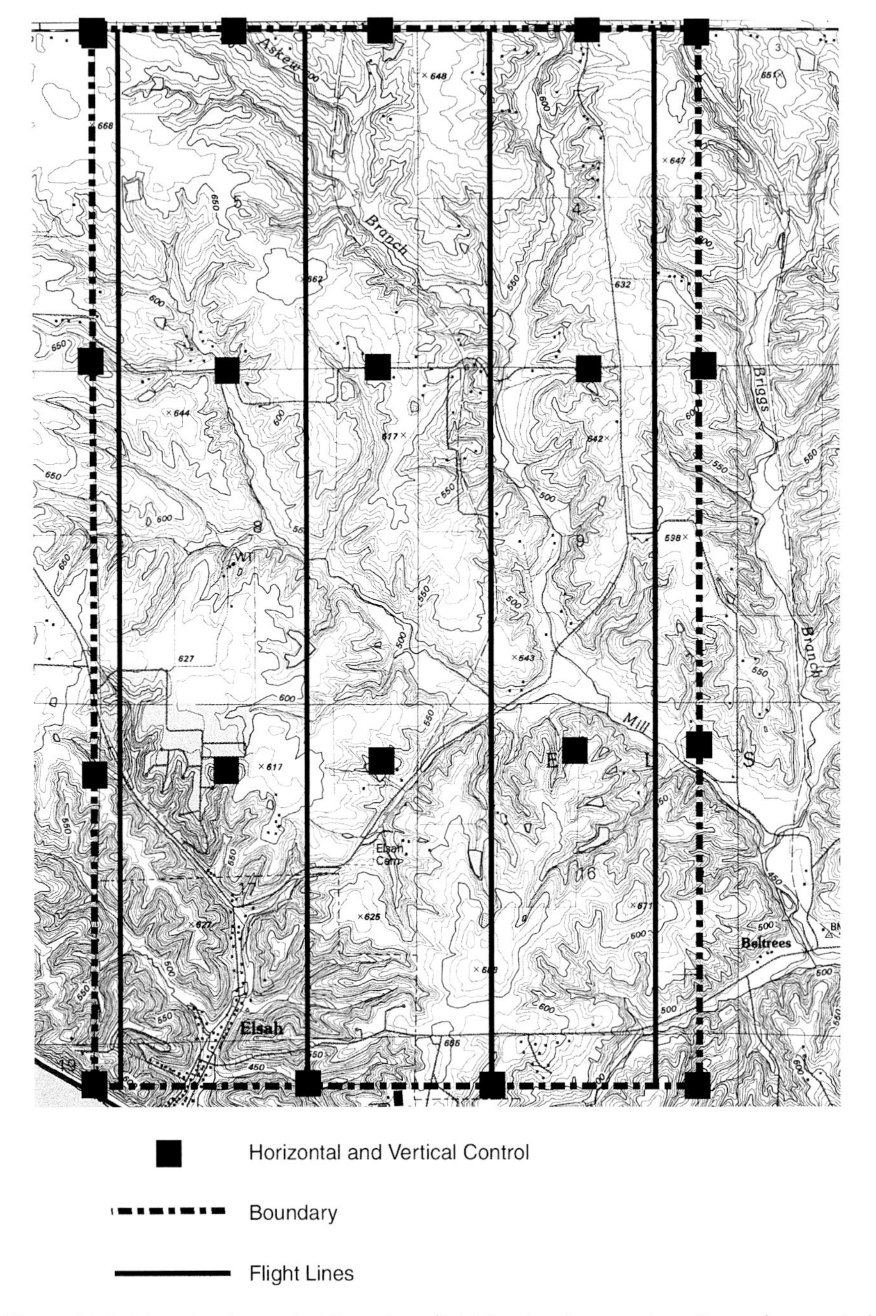

Figure 11.4 Map showing project boundary, flight line locations, and positions of suggested control point locations.

CHAPTER 12

Photogrammetric Map Compilation

12.1 DIGITAL MAPPING DATA

The latter part of the 20th century brought an insatiable demand for georeferenced data. Not only does the business world now demand that information and statistics be geographically referenced, but the scientific and engineering industries do as well. Most digital data derived from aerial photography and other remote sensors will eventually find its way into georeferenced databases. Data from these and other sources can be earth registered, mixed and matched, and incorporated into Geographic Information Systems (GIS) for detailed analysis and solutions. Therefore, it is imperative for the project manager to understand the implications of data and coordinate systems, as well as methods of map data compilation.

12.2 COORDINATE SYSTEMS AND DATA

The demand for georeferenced data today is constant and touches all aspects of our daily life. It is therefore important to consider the potential uses of data sets and their respective coordinate systems and data. These choices affect the utility and accuracy of the final data set.

12.2.1 Coordinate Reference Systems

A coordinate reference system is an important consideration in a mapping project since it is a major factor in the accuracy of the mapping data set and should be specified with the chosen horizontal and vertical data.

North American Datum of 1927 (NAD 27) and National Geodetic Vertical Datum of 1929 (NGVD 29) are data that have been used throughout the United States for many years. They are the basis for many historical mapping data sets and are therefore still required for certain projects. The NAD 83 and North American Vertical Datum of 1988 (NAVD 88) are the most current horizontal and vertical reference datums used in the United States.

Conversion techniques between coordinate reference systems are mathematically feasible. Since this procedure may affect accuracy, it should be accomplished only by a fully qualified professional (i.e., geodisist or registered land surveyor).

12.2.2 Coordinate Systems

Data are referenced horizontally to one of several grid coordinate projection systems. Coordinates can be referenced to any desired geographic system, and the selected system is at the discretion of the user. This can be either a standard reference system to tie the mapping to the world or the state or an assumed grid reference pertinent only to the specific site.

It would be best to consider using standard coordinate systems for geographically registering information to facilitate comprehensive use of the data. This chapter will discuss three of the more universal grid systems accepted and commonly used in the United States. Selection of a datum for a project should consider end-user demands and accuracy.

An ideal situation would have an entire project in one segment of a selected datum; however, some projects do not lend themselves to this situation. Projects that involve an area that runs north/south for a long distance may cross datum segments. A common resolution to this problem can be to select one datum segment and extend it through the entire project. Obviously, this can cause horizontal accuracy issues that must be considered and understood. These issues should be discussed with a professional individual well versed in each datum and conversion.

12.2.2.1 Universal Transverse Mercator

UTM is a worldwide planar map projection that breaks the globe into 60 zones, each covering 6 degrees of longitude. UTM is a cylindrical envelope intersecting the earth along two lines that are parallel to a central meridian. This datum is commonly used for projects covering large areas. UTM grid structures are generally expressed in metric units and are tied to a reference datum (i.e., NAD 83). The United States is covered by UTM zones 10–20.

12.2.2.2 State Plane

The State Plane Coordinate System (SPCS) segregates 120 zones throughout the United States. Conformal conic projection can be used in states that are wide in the east–west direction. Transverse mercator projection can be used in states that are narrow in the east–west direction. These data are commonly used for projects falling within the boundaries of a specific state. The coordinates reference a specific SPCS along with an X and Y position in a plane and are expressed as "easting" and "northing," respectively. The units may be expressed in United States feet or metric units.

SPCS-referenced projects that extend across state boundaries are sometimes necessary. When this situation is encountered, a common resolution is to choose one SPCS and zone and extend it throughout the entire project. In this situation horizontal errors will occur and should be understood; however, these errors are generally relatively small and can usually be accepted.

12.2.2.3 Latitude/Longitude

Latitude/longitude coordinates are not normally used on large-scale mapping projects, but they are common to small-scale projects covering large areas. If captured data are to be used in a national or international database, then this projection may be appropriate.

12.2.3 Vertical Data

The design of a mapping data set should always consider the vertical as well as the horizontal data, and accurate mapping databases must also be referenced to vertical data. Horizontal accuracy is affected by vertical accuracy. Therefore, the selection of the vertical datum is just as important as the selection of the horizontal datum. These vertical data are tied to a network of established ground points commonly called benchmarks.

NGVD 29 and NAVD 88 are the two most common vertical data used in the United States. A collection of benchmarks for NGVD 29 began in the 1850s and was first established in 1929 by the U.S. Coast and Geodetic Survey (USC&GS). Over time, the earth began to move and adjust, surveying methods changed and became more accurate (i.e., GPS), and NGVD points were continually obliterated by urban progress. A new set of points was established by the NGS, beginning in the late 1970s and completed in 1991. This new vertical datum is known as NAVD 88.

Adjustments have been made to NAVD since its inception, and care should be taken to note the desired adjustment for a specific mapping data set.

12.3 STRUCTURE OF DIGITAL DATA

Digital information that represents a map resides in a database matrix composed of a multitude of individual points. Each point is spatial, having a coordinate triplet value for X (easting), Y (northing), and Z (elevation).

12.3.1 Digital Data Generation

Mapping computer software does not "know" whether it is producing a contour or a building. It captures points and draws lines. To plot a map for human viewing, map features created by digital data generation must be described by a composite of several characteristics. Figure 12.1 schematically depicts the path of spatial information that is collected by a stereoplotter:

- Spatial (XYZ) data points, which form planimetric or topographic features, are generated by the stereoplotter.
- Generated data are stored within the data collector.

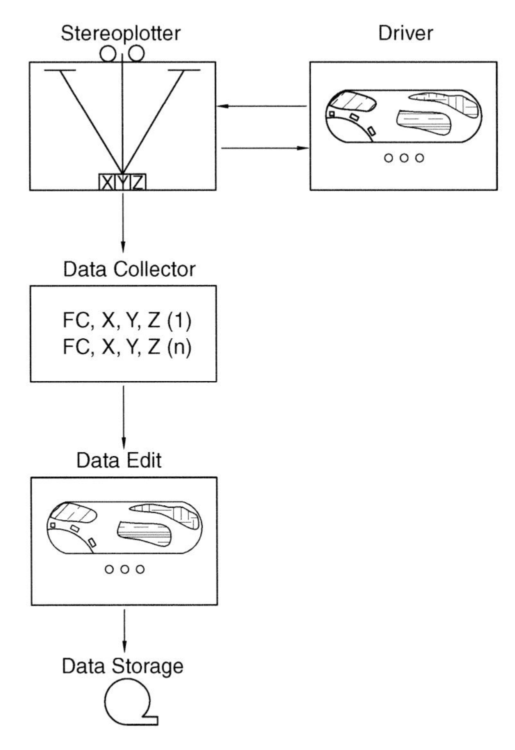

Figure 12.1 Schematic rendition of the path of spatial information that is collected by a stereoplotter.

- The created features appear on a monitor, in the form of mapping symbols, which is within view of the technician.
- Mapping data are output to peripheral editing, where a technician makes corrections to the database.
- Data can be output as hardcopy maps or on various electronic data media.

12.3.1.1 Feature Code

A feature code (FC) is an alpha/numeric character set that the computer recognizes as an identifier of a feature. This code directs the computer to an appropriate macro file.

12.3.1.2 Macro File

A macro file is composed of a set of instructions that controls the characteristics of a feature symbol. These instructions can pertain to line style, weight, and color, as well as any other specific symbol characteristics that are necessary to create the map feature.

12.3.1.3 Data String

A data string is a consecutive series of coordinate triplets (XYZ) which guides the computer in placing the feature in its true geographic location.

12.3.1.4 Data Form

This protocol readily permits differentiation of features when viewing the graphics screen or a data plot. The computer treats digital data as:

```
FC, XYZ(1),.....,XYZ(n)
```

12.3.2 Automated Feature Collection Methods

Advancements in software have changed the way data sets are collected. The project manager must be familiar with and understand the difference and importance of several other key terms:

1. *Cells* are planimetric features that have a standard shape and are repeated frequently throughout a mapping data set. They can be very intricate (i.e., a specific type of equipment) or very general (i.e., a utility pole). Mapping software can configure these cells into general groups sometimes known as cell libraries.
2. *Mass points* are points horizontally and vertically established on the mapping surface that depict a change in the vertical character of the surface (i.e., high or low spot on the ground).
3. *Breaklines* are unique, linear, topographic features that depict an abrupt change in the mapping surface (i.e., drainage patterns, edge of walls or roads, or cliffs). Breaklines define nonuniform character in the mapping surface.
4. The *Digital Elevation Model* (DEM) is a collection of evenly spaced elevation points that depict the mapping surface.
5. The *Digital Terrain Model* (DTM) is a collection of mass points and breaklines.
6. The *Triangulated Irregular Network* (TIN) is a surface model of triangular shapes created by the collection of DEM and DTM data.
7. *Contours* are lines of equal elevation on the mapping surface.

12.3.3 Data Collection

Data collection can generally be divided into three basic categories.

12.3.3.1 Planimetric Features

Cultural details are generally considered to be structures, road networks, edges of water bodies, utilities, transportation facilities, etc.

12.3.3.2 Topographic Features

Terrain characteristics are considered to be changes in the elevation of the Earth's surface.

12.3.3.3 Annotation

Attributions include textural information that describes the planimetric or topographic features. The planimetric feature annotation may include street names, building numbers, or the type of construction. The topographic feature attribution may include the horizontal and vertical location of the feature (i.e., elevation 510 ft).

12.4 ADVANCEMENTS IN MAP COMPILATION

Current advancements in map compilation software have improved the efficiency and accuracy of feature data collection.

Linear feature recognition software is capable of recognizing and automatically plotting certain planimetric features such as road networks and edges of water bodies. The success of these types of software depends largely on the quality of the source imagery and the desired features to be collected.

The collection of common, often repeatable planimetric features (i.e., certain building types or utilities) with the use of cells and cell libraries can speed the process and in some cases eliminate errors.

Software that makes use of DEM, DTM, and TIN data sets to generate cross-sections and contours also can expedite data collection and minimize errors.

12.4.1 Elevation Data Collection Methods

Elevation data are generally collected as a combination of mass points and breaklines and is processed through specific software to generate data sets such as TIN and contours. Today, end users of elevation information utilize data sets for analysis and design.

See Figure 12.2 for a scheme of mass points and breaklines.

1. Mass points can be created in one of two methods:
   a. A pattern of points, either random or on a grid, can be read on a stereoplotter or softcopy.
   b. A process of autocorrelation can be implemented so that the computer generates a grid of spatial points.
2. Breaklines are strings of individual spatial points read on a stereoplotter or softcopy following abrupt terrain change features.

Figure 12.3 depicts the contours created using the mass points and breaklines seen in Figure 12.2.

12.4.2 Planimetric Feature Collection Methods*

Planimetric features such as buildings, utilities, roads, water body boundaries, and edges of tree lines are collected by viewing imagery and collecting a series of points

* Chapter 14 in *Aerial Mapping: Methods and Application* (Lewis Publishers, Boca Raton, FL, 1995) presents a broader version of data collection.

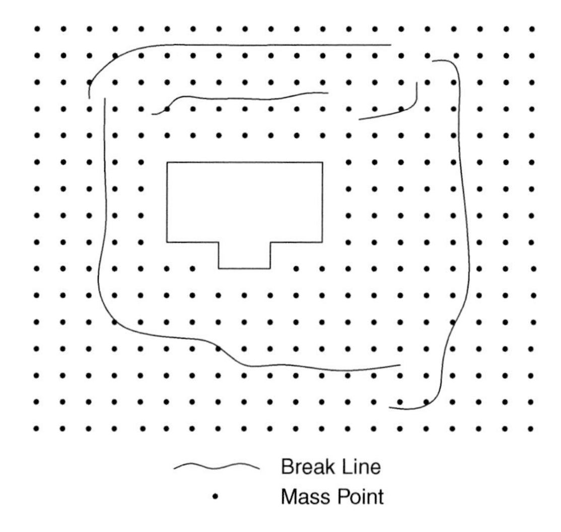

Figure 12.2 A scheme of mass points and breaklines.

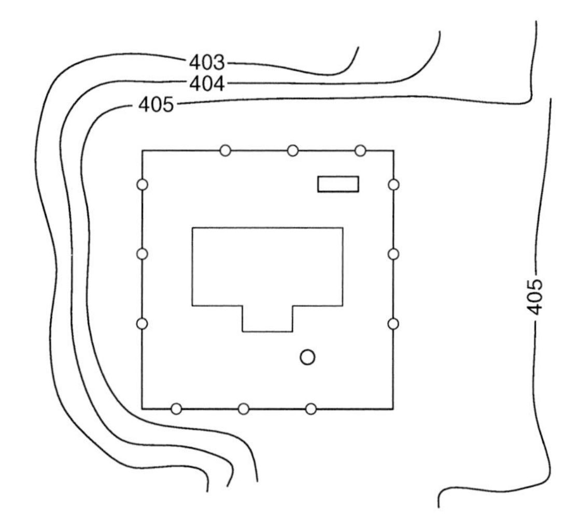

Figure 12.3 Contours created from mass points and breaklines.

that describe the location and shape of the feature. The data point stream that delimits the shape of the feature is accompanied by factors that identify it — line weight, style, and color, as well as any other factors that further characterize the feature.

Many of these planimetric features can also be collected with the aid of linear feature recognition software. This technology is still under various stages of development, and its success depends to a great degree on very high quality imagery in order to distinguish selected features to be collected. Feature recognition software requires the use of digital imagery and softcopy compilation. Generally, the process involves the selection of a typical group of pixels that represent a feature (i.e., a paved road surface). The software is then directed to follow a certain pattern and find pixels that are the same as those selected. The software then generates a prescribed pattern for the feature noted.

12.5 DATA STANDARDS

Feature standards have been developed and adopted by most major mapping firms and by federal and state agencies. Features of constant shape, called cells, were increasingly used in mapping features such as fence and tree lines, utility poles, and railroad tracks. Standard line weights, colors, and patterns were developed. The use of geospatial data in virtually every part of our lives at home and work has demanded a common set of standards for spatial data. The federal government has developed the Federal Geospatial Data Standards, and these standards are becoming the default standards for the photogrammetric mapping industry. A successful project manager will include the appropriate set of mapping standards in the scope of work for any photogrammetric mapping project.

12.6 DIGITAL MAPPING DATA FLOW

Successful project managers of photogrammetric mapping projects need not comprehend the mathematical equations that are the basis for stereodigital data collection. However, they should have a general understanding of the basic principles and work flow. Figure 12.4 shows a general stereoimage collection work flow chart.

12.6.1 Project Planning

A successful stereocompilation project should begin with a set of goals, including the expected end products, level of detail required, standards, horizontal and vertical accuracy, and final data media. These goals can be achieved by writing an adequate scope of work that clearly defines the end results, but does not dictate methodology:

- End products
  - Planimetric and topographic maps
- Level of detail required
  - 1 in. = 100 ft with 2-ft contours
- Standards
  - FGDC spatial data standards
- Accuracy requirements
  - ASPRS Class 1 standards
- Final media
  - Digital data on CDROM
  - Hardcopy on bond paper

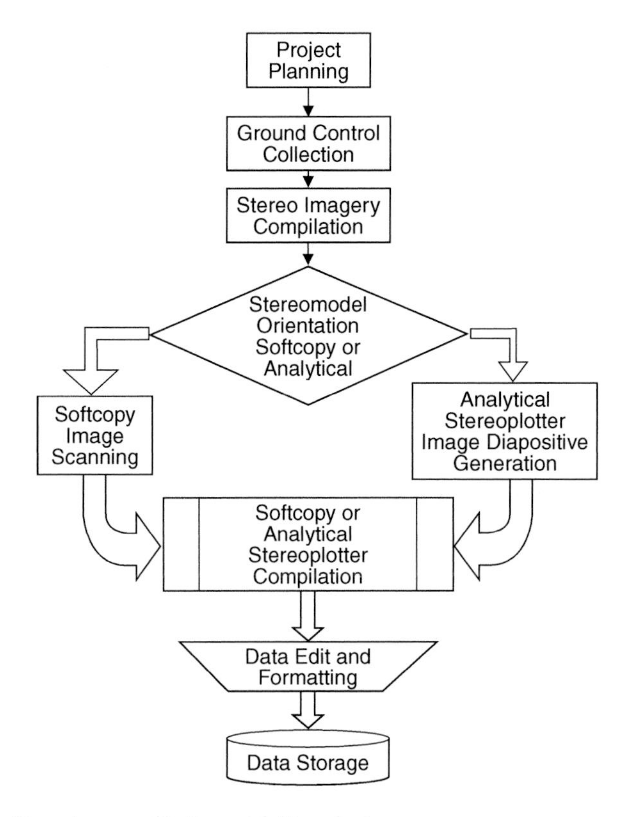

Figure 12.4 Stereoimage collection work flow chart.

12.6.2 Ground Control Collection

Ground control installation should be accomplished by methods that meet the accuracy requirements for the final mapping products only. Survey efforts such as permanent retrievable control points for many mapping projects may not be required and generally should not be specified in planning and scope of work development. The amount of ground control required and its accuracy are tied to the mapping accuracy requirements. Therefore, the methods used for ground survey collection, number and location of points, etc. may not need to be specified. These details should be left to qualified professionals such as the photogrammetry firm professionals who will accomplish the mapping.

12.6.3 Imagery Collection

For the purpose of this chapter, imagery collection will be assumed as utilizing new near-vertical aerial photography. However, some projects may allow for the purchase of existing photography or other sensor type data to be used for the final mapping data set. Imagery type, camera lens focal length, negative scale, and coverage should be clearly understood:

- Imagery type
  - Black and white
  - Natural color
  - Color infrared
- Camera and photo scale
  - Flown with a 6-in. focal length metric camera
  - Near-vertical photography at 1 in. = 300 ft negative scale
  - Flow with 60% forward lap and 30% sidelap
- Photo coverage
  - Photo coverage as shown on USGS 7.5 ft quad sheets

12.6.4 Stereomodel Orientation

The next major phase in the mapping process is orientation of the imagery to the Earth's surface by using the ground control collected for the project. This process is most often accomplished today with the use of aerotriangulation software (see Chapter 11 for more detail).

Stereomodel orientation consists of:

- Adjustment of imagery to correct camera alignment to see in three dimensions
- Adjustment of the imagery to the earth's surface
- Adjustment of the imagery to ground control

12.6.5 Digital Data Stereocompilation

Currently, the mapping industry is moving toward a total softcopy mapping environment. Softcopy mapping system hardware is generally less costly than the conventional analytical stereoplotters of the 1990s. The efficiency of software coupled with the increased speed of microprocessors has made the feature collection accuracy and speed comparable with that of conventional analytical stereoplotters. When capturing digital data with an analytical stereoplotter, the mapping instrument is interfaced with a driver computer that controls the actions of the mapping instrument. Digital data are then exported to a data collector.

Softcopy stereoplotting systems incorporate a high-resolution scanner, a large format/high-resolution monitor, a high-speed computer processor, and large data storage media with software capable of viewing digital imagery in three dimensions with the aid of special glasses. A suite of software allows for the viewing and compiling of features from digital imagery provided from high-resolution scans of original processed aerial film.

Digital mapping with an analytical stereoplotter or softcopy workstation involves a series of operations for both planimetric and topographic feature collection.

12.6.5.1 Planimetric Features

Planimetric (cultural) data are generated by drawing cultural features that are visible and identifiable on the image. Within the stereoplotter there is a visible reference mark. To draw an object the operator must position this reference mark on the apparent elevation of that feature. Since mapping is accomplished by utilizing

Table 12.1 Common Planimetric Features which Are Present in a Digital Database

Airports	Footpaths	Railroads
Alleys	Golf courses	Reservoirs
Athletic fields	Guardrails	Rivers
Barriers	Hydrants	Roads
Billboards	Inlets	Ruins
Borrow pits	Jetties	Sidewalks
Bridges	Lakes	Sign posts
Buildings	Light standards	Single trees
Canals	Mail boxes	Streams
Catch basins	Manholes	Streets
Cemeteries	Orchards	Substations
Churches	Parking lots	Swamps
Dams	Parks	Tanks
Ditches	Piers	Telephone poles
Driveways	Pipelines	Towers
Dwellings	Plantations	Trails
Fences	Pools	Walls
Field lines	Power structures	Wharves
Flag poles	Quarries	Woodlands

a spatial image, the mark must be at its true elevation for the observed point to be in its true horizontal position.

Unless otherwise specified, compilers map all planimetric features that are compatible with the mapping scale. On large-scale photos, various cultural features can be identified. A list of these features would include, but would not be limited to, those tabulated in Table 12.1. The compilation technician must also know and utilize the symbology and scheme that is called out in the scope of work for a project:

- Plot all features visible and identifiable on the imagery
  - Depends upon the imagery scale and project specifications
- Symbology and scheme
  - Specifications should note required symbology
  - Feature layers required (scheme) should be clearly spelled out in the scope of work

12.6.5.2 Topographic Features

Data necessary to create contours are collected by the process of DTM and by TIN generation. Contour intervals should be known and addressed in the scope of work prior to collection. Selected mass points are also collected to model surface area that is not typical (i.e., high and low areas). These types of points are placed individually by the compilation technician and are therefore generally more accurate than computer-generated points (DEM) and contour lines.

To obtain contour and elevation data:

- Automatically generate a DEM
- Generate mass points and breaklines where necessary (DTM)
- Review stereomodels and add any required additional points and breaklines to define irregular areas

12.6.5.3 Data Scheme

Digital data are usually separated into "layers," with each data layer containing specific information for related features. Today this is generally called a data scheme. Many data sets are used in a GIS or are stored for use at some later time. The data scheme greatly affects the effectiveness of a data set in a GIS. Individual layers contain information concerning contours, transportation, hydrology, photo control point locations, map compilation boundaries, vegetation, transmission lines, surface utilities, underground utilities, political boundaries, buildings, text annotations, and other information essential to fulfilling the demands of a query. The scheme to be used should be specified in the scope of work.

The federal government has developed a broad set of data scheme known as the Federal Geodetic Data Committee Spatial Data Standards (FGDC-SDS). These standards are available to the public and have been developed with experts in the mapping industry. In lieu of other unique project scheme standards, FGDC-SDS may be employed.

12.6.5.4 Digital Data Edit

In a state-of-the-art photogrammetric mapping office today, editing is generally divided into two parts:

- Quality control of the mapping products. Most analytical and softcopy workstations today are capable of viewing feature vector data (i.e., roads, structures, contours, elevation points, etc.) overlaying the imagery. This software feature allows for a thorough review of the data collected. One method of checking elevation data is accomplished by generating contours from the elevation points collected. The contours are then superimposed over the imagery, and the editing technician checks to make sure that they accurately depict the character of the ground and are attributed properly. Planimetric feature data collection quality control is also best checked with the aid of superimposition over the imagery.
- Graphic editing and final formatting of the data sets. Graphic editing is generally considered to be the final attribution of files, sheeting, and formatting. Attribution may include margin and legend information, tic and grid marks, building and street names, and contour labels. Many projects will always require hardcopy maps with preset sheet formats, borders, and title blocks. The graphic editing technician prepares files for these types of products and formats the digital data sets in the required software format, such as Intergraph, Arc Info, or AutoCAD. The graphic editing technician then creates the final hard- and softcopy data sets for delivery.

Data Editing
Quality control checks
Planimetric data
Topographic data
Graphic editing
Attribution
Sheeting
Final data formatting

12.6.6 Ancillary Data Collection

The demand for spatial data today has also increased the necessity for secondary spatial data. A secondary spatial data set contains information that is generated from the primary planimetric and/or topographic data sets. Many engineering, planning, and environmental projects demand the capability of geospatially accurate planimetric and topographic data to provide answers to queries. Volumes, centroids, areas, and linear distances are all easily calculated from geospatially accurate data sets. Today, software packages allow the photogrammetric technicians to compile these types of data as they collect the features. When these demands are known at the beginning of a project, the effort to collect the information and the associated cost can be minimized. Again, spending the time to create an accurate set of specifications that clearly spells out the required data sets is imperative to the project manager.

12.6.7 U.S. Geological Survey Data

The USGS stocks a number of products available to the public.

Millions of aerial photos within the confines of the United States and images covering the world from several series of remote sensing satellites are stored at the EROS Data Center. Queries can be directed to:

U.S. Geological Survey
EROS Data Center
47914 252nd Street
Sioux Falls, SD 57198-0001
Tel: 800-252-4547 or 605-594-6151

A list of US GeoData products — digital line graphs (DLG), digital orthophoto quads (DOQ), digital raster graphs (DRG), and digital elevation models (DEM) — is located by an on-line search of the key phrase "global land information."

USGS quadrangle sheets of various scales can be purchased from numerous dealers or from the Rocky Mountain Mapping Center. Log on to the web site http://mapping.usgs.gov/mac/maplists.html for a list of the obtainable maps. This reference tells how to locate map sheets by state or specific sheet name or by latitude/longitude, and is also a guide to local distributors.

CHAPTER 13

Information Systems

13.1 INFORMATION SYSTEMS

Significant amounts of data that are collected through digital aerial mapping, remote sensing, and image analysis are stored in databases for information systems. There are several types of information systems, including Geographic Information Systems (GIS), Land Information Systems (LIS), and Automated Mapping/Facilities Management (AM/FM). Not all information systems serve exactly the same purpose, but most have similar functions.

This chapter will touch briefly upon GIS. According to ASPRS, a GIS "is an information system able to encode, store, transform, analyze, and display spatial data."

The concept of GIS is over 100 years old. Originally, information layers were represented by individual transparent or translucent maps that were overlain one atop another to form a physical composite of data layers. Early linear structured computers, along with rudimentary peripheral input and output devices, were unsupportive of the demands of GIS functions. Beginning in the late 1960s, computer-based GIS became a working reality. Current working environments for a GIS would be CADD/CAM/CAD or softcopy.

13.1.1 Value of Geographic Information Systems

The primary advantage of a GIS is the capability of the system to reach into the database and select particular portions of information necessary to formulate alternative solutions to given scenarios.

13.1.2 Demands of a Geographic Information System

A GIS is a very demanding information collection and manipulation environment incorporating a number of characteristics:

- The user must be permitted to constantly interact, issuing instructions and receiving responses, with the computer system.
- Output from GIS systems must be graphic. Such output devices would include high-resolution color graphic screens and data plotters.

- Geographic information sets are large and complex, requiring huge amounts of digital storage to be readily available.
- Hardware engines must be capable of processing large volumes of layered information quickly and providing timely solutions to queries.
- Computer hardware configurations and software packages are necessarily complex to accommodate the multidimensional (XY, XYZ, attribute) nature of geographic data. Therefore, the database management systems (DBMS) must be comprehensive and sophisticated.
- Benefits derived from the application of GIS must exceed the cost of these environments. Fortunately, the cost of increasingly more powerful hardware and efficient software continues to lessen. Hence, these facilities are available to a varied market.
- Information from a multitude of diverse sources can be integrated into a database. Information systems must be capable of comparing blocks of dissimilar data and presenting the viewer with a composite scenario based on given situation parameters. This allows the manager to manipulate variable parameters to compare multiple solutions with a limited expenditure of time and manual effort.

13.2 COMPUTER-AIDED MAPPING

Map compilation and analysis are accomplished today through the use of specialized computer software algorithms designed to perform engineering drafting and design (CADD) and GIS functions. CADD software is usually designed to be more robust in the compilation of feature detail, while GIS software usually provides more spatial analysis tools. It is rare today for a mapping project to only demand two-dimensional mapping data sets without real spatial ties. Therefore, current demands for spatial data compilations and analysis usually require that map compilation systems include both CADD and GIS capability used in tandem.

Software vendors such as Intergraph, AutoCAD, and ARC/INFO are major developers of map compilation and spatial analysis tools. Software packages supplied by these and other vendors provide a suite of software tools that are capable of map feature compilaton and GIS analysis. Another critical tool in these packages is digital image viewing and scaling tools which allow the mapping scientist to compile feature detail, annotate features, geographically locate features, and store the information in databases for ease of analysis.

13.3 THEMES

A theme encompasses an area of similar characteristics. Collected data for various themes are placed in specific data layers for convenience in evaluating the database. Individual layers must be georeferenced to the ground using a common grid system, such as the state plane, UTM, or latitude/longitude. This assures that the data from various layers geographically matches when composited.

13.4 DATA COLLECTION FOR INFORMATION SYSTEMS

A GIS database is integrated with information from diverse sources to propagate information system solutions and products. Data from these sources can be mixed and matched in analyses ventures. Data derived from remote sensors other than aerial cameras are incorporated into databases. As a consequence, photogrammetrists, especially those who operate softcopy systems, should have some knowledge of remote sensors, image analysis, and GIS to correctly apply these technologies.

Except for the most uncomplicated GIS studies, the expertise of several technicians proficient in various fields of discipline may be required to affect a reliable solution.

13.5 U.S. GEOLOGICAL SURVEY INFORMATION SOURCES

The USGS is a source of information which GIS project managers may wish to look into.

13.5.1 Tutorial

The USGS furnishes an illuminating tutorial describing the concept of a GIS at the web site http://www.usgs.gov/research/gis/title.html, explaining what a GIS is and how it works. Its step-by-step progression with easily understood text and complementary illustrations delves into the following:

- Relating information from different sources
- Data capture
- Data integration
- Projection and registration
- Data structure
- Data modeling
- Information retrieval
- Topological modeling
- Networks
- Overlays
- Data output
- Applications

13.5.2 Geospatial Information

The USGS is also a source of obtainable geospatial information that a project manager may want to examine as a source of digital mapping products, which may be especially useful for GIS/LIS projects on large areas. Log on to the web site http://mapping.usgs.gov/www/products/status.html to learn, by state, the status of such geospatial data products as:

DEM (digital elevation models)
DOQ (digital orthphoto quads)
DLG (digital line graphs) with specific overlays
DRG (digital raster graphs)
NAPP (national aerial photographic program)

A list of USGS National Mapping Program geospatial data and products that are available can be found on the web site http://mapping.usgs.gov/www/products/1product.html.

The USGS web site http://rmmcweb.cr.usgs.gov/elevation/dpi_dem.html describes DEMs and how to order this data media which can be employed to create contours and orthophotos.

Technicians who are planning to interject USGS layered data into GIS projects may want to access the web site http://rmmcweb.cr.usgs.gov/nmpstds/dlgstds.html for information regarding the National Mapping Program Standards for 1:24,000 scale DLG data layers involving hydrography, transportation, boundaries, public land survey system, built-up areas, hypsography, vegetative surface cover, and named landforms.

A listing of available individual 1:100,000 scale DLGs that are available within the confines of various states can be located by logging on to http://edcwww.cr.usgs.gov/glis/hyper/guide/100kdigfig/states/??.html. To access a particular state, substitute the two-letter state code for ?? in the web address.

A DRG is a scanned image of a standard series topographic map referenced to a UTM grid. The web site http://mcmcweb.er.usgs.gov/drg/ cites information about the availability of this raster data.

Information regarding multidisciplinarian GIS projects specific to the sciences of biology, geology, mapping, and hydrology can be acquired through the auspices of the USGS. A search of the web site http://ask.usgs.gov/sources.html will lead to authorities throughout the United States, specialists located in field offices, facilities specializing in pertinent sciences, and partnerships with other organizations. This source furnishes lists for:

- Earth information centers
- State water resources representatives
- Libraries for biology, geology, mapping, water
- USGS libraries
- Scientific partnerships
- USGS offices

13.5.3 Areas of Specialty

Information paths can also be traced to areas of specialties, such as:

- Earth Observation System Data Center
- National Earthquake Information Center
- National Landslide Information Center
- National Geomagnetic Information Center

- Coastal and marine geology
- Energy sources
- Minerals information
- Biology and ecology centers
- Volcano observatories

CHAPTER 14

Orthophotography

14.1 GENERAL

Today, many uses for geospatial mapping products require current planimetric feature data. Analysis and design from geospatial data sets generally require a known positional accuracy of features. The collection and updating of planimetric features in a data set can be costly. Many end users are also not accustomed to viewing and analyzing vector-based mapping data sets. They prefer to view planimetric features as a photo image. As an example, the Arch National Monument in St. Louis, MO would simply be drawn as a long, narrow strip vector shape parallel to the Mississippi River on a vector map. An orthophoto map would show the arch as an easily identifiable unique image feature.

The cost to collect and update planimetric features can be significant. Costs can sometimes be minimized by the production of a photo-based digital map set that is spatially accurate throughout. Many GIS data sets make use of photo-based image data for these purposes.

The Internet offers orthophotograph tutorials which may aid managers interested in the basic study of the subject. Those interested can refer to the following web sites:

- One which comments on principles, project design, issues, utility, accuracy, and economics can be found at http://www.gisqatar.org.qa/conf97/links/h1.html.
- The other — dealing with aerial photography, scales, relief displacement, digital orthophoto generation, accuracy, and image quality — can be found at http://pasture.ecn.purdue.edu/~aggrass/esri95/to150/p124.html.

14.2 ORTHOPHOTOS

Orthophotographs are photographic images constructed from vertical or near-vertical aerial photographs. The processes used to generate orthophotos remove the effects of terrain relief displacement and tilt of the aircraft. When properly generated, these digital images have a predictable constant positional accuracy throughout the entire image (see Figure 14.1).

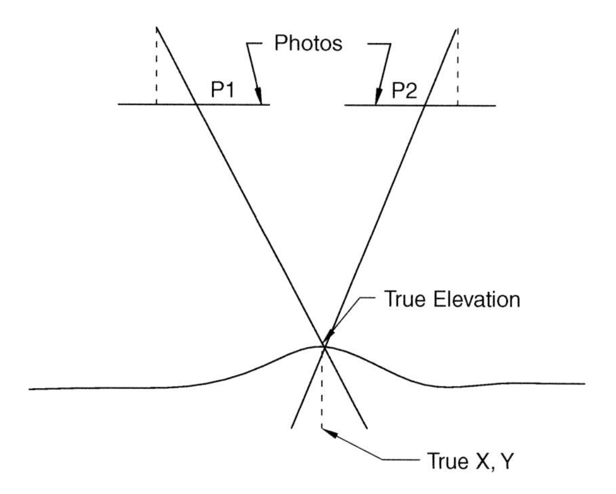

Figure 14.1 The effects of relief and how it is corrected for orthophotos.

A digital orthophoto image is a raw digital aerial photo image rectified to a suitable DEM of the same area. Software merges the digital image with the DEM and aligns the image orthogonally.

Currently, scanning technology and software allow end users to generate products that may appear as orthorectified images at very little cost. However, the techniques used to generate many of these products will not produce a positionally accurate digital image. The employment of such products may be justified for some projects, but should never be confused with or considered as a digital orthophoto. Nondigital orthophoto products are generally considered to be digital image enlargements or semi-rectified digital images.

14.3 DIGITAL ORTHOPHOTO IMAGE PRODUCTION

Figure 14.2 illustrates a flow diagram of orthophoto production.

14.3.1 General

Many end users of digital orthophotos today have very robust hardware and sophisticated software to view and manipulate orthophoto images. They require the ability to view selected image features and perform analysis such as relative distance, area, or even change analysis. In order to meet these demands, proper design of an orthophoto is imperative. Design should consider the following factors:

1. Expected uses of the orthophotos and smallest features to be viewed and studied
2. Accuracy requirements (relative and feature)
3. Anticipated equipment with which the orthophotos will be viewed
4. The equipment, data, and processes used to generate the orthophotos

The end user and project manager should consider and be prepared to relay the information in Items 1, 2, and 3 to the photogrammetry technician. The photogrammetry

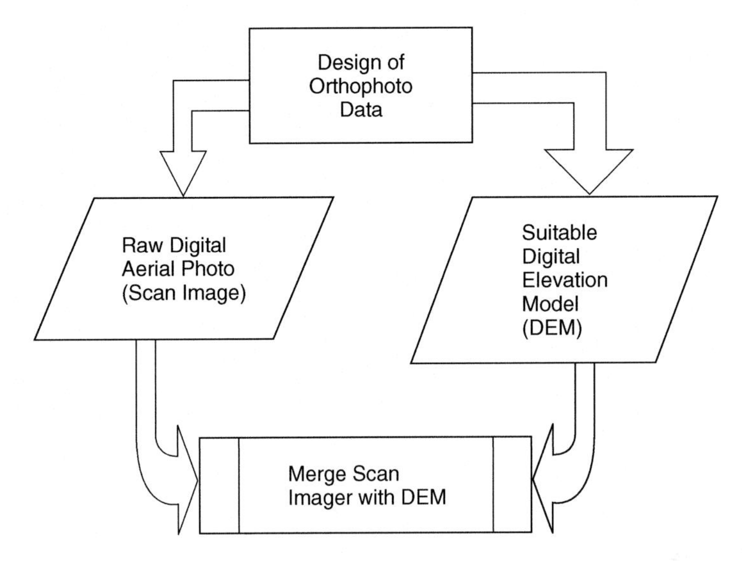

Figure 14.2 Flow diagram of orthophoto production.

technician should then design the orthophoto data collection and production around the equipment and processes necessary to meet these requirements (Item 4).

14.3.2 Design Parameters

Design parameters for an orthophoto are generally tied to the expected final accuracy. Suitable imagery and ground control are the basic elemental data that determines the final orthophoto reliability which involves both the accuracy of distances and areas within the orthophoto as well as the relative accuracy of features with respect to their true location on the earth. Distance and area accuracy are based on the pixel size. Relative feature precision is based on the accuracy of the DEM used in the rectification process. The relative accuracy cannot be more precise than the reliability of the DEM.

14.3.2.1 Imagery and Ground Control

Proper selection of imagery scale and ground control as stated above are critical to the reliability of the final orthophoto. Imagery either can be from existing sources (aerial photography or satellite/airborne imagery) or can be obtained specifically for the project. The key is the suitability of the imagery to meet the intended uses of the orthophotos. Some items to consider are as follows:

- Scale of the imagery
- Type of imagery required (i.e., black and white, natural color, and color infrared)
- Clarity of the imagery (i.e., cloud cover, vegetation cover, seasonal requirements)
- Timeliness of the imagery
- What is the format of the imagery and how effectively can it be introduced into the orthophoto generation process

Table 14.1 Digital Orthophoto Enlargement Factor from Photo Negative Scale

ASPRS Class	Enlargement
1	4–7 times
2	6–8 times
3	7–10 times

A frequently accepted accuracy standard for photogrammetric mapping and orthophotos is the ASPRS standards that have been discussed in detail in Chapter 8. Imagery design for orthophotos can therefore be tied to the ASPRS Class 1, 2, and 3 horizontal accuracy requirements and an expected ground pixel size (resolution). Generally, the pixel size is based on the scale of the photonegative. For example, if the smallest feature that the end user must be able to see in an orthophoto is a typical sewer manhole (approximately 2 ft in diameter), then the imagery must be of a horizontal scale capable of viewing the feature and the pixel resolution should be approximately 2 ft.

Photogrammetric equipment today allows for suitable orthophotos to be generated from photography at photo negative scales that are smaller than the intended final orthophoto scale. Table 14.1 lists recommended digital orthophoto enlargement factors for photo negative scale.

To illustrate, a final 1:2400 (1 in. = 200 ft) scale orthophoto that requires ASPRS Class 1 horizontal accuracy may be obtained from an aerial photo scale of 1:7200. A 1 in. = 200 ft photograph will be able to adequately produce a 1-ft pixel resolution image. This capability allows for time and cost savings in the production process.

14.3.2.2 Image Scanning

Imagery for digital orthophotos may be converted to a digital image with a specific pixel resolution. Obviously, existing digital imagery that is suitable for an orthophoto image does not require this process. A suitable digital image from original processed analog film may be created with the aid of a high-resolution metric scanner such as that illustrated in Chapter 7, Figure 7.5.

These scanners are capable of scanning processed aerial film at very high resolutions (7 μm) if required. High-resolution scanners also have the capability of scanning black and white, natural color, and color infrared film. The end user must be mindful of the fact that color digital images require three times the data storage space. This fact can affect performance and utility of the final products and should be taken into consideration during the design of an orthophoto project. Each pixel consists of a radiometric value plus an XY coordinate set. Radiometric gray scale of a single picture element may fall between reflectance values 0–255. Zero is no reflectance (black), and 255 is full reflectance (white).

Both quality and economy must be factored into the selection of the pixel size. As discussed earlier in this chapter, proper flight altitude and scan rate must be designed for the orthophoto design horizontal scale. Reducing pixel size greatly increases database magnitude, which affects storage capacity and processing time.

Table 14.2 Recommended Approximate Pixel Sizes for Selected Digital Orthophotograph Map Plot Scales

Final Map Plot Scale	Approximate Ground Pixel Resolution Required to Meet ASPRS Accuracy Standards
1:500	0.0625 m
1 in. = 50ft	0.25 ft
1:1000	0.125 m
1 in. = 100 ft	0.5 ft
1:1500	0.250 m
1:2000	0.375 m
1 in. = 200 ft	1.0 ft
1:2500	0.5 m
1 in. = 400 ft	2.0 ft
1 in. = 500 ft	2.5 ft
1 in. = 1000 ft	5.0 ft

Table 14.3 Digital Orthophoto File Size Based on a Neat Double (7.2 × 6.3 in.) Model for Black and White Uncompressed Images

Scan sample rate	7.5 μm	15 μm	22.5 μm	30 μm
	3386 dpi	1693 dpi	1128 dpi	846 dpi
File size	496 Mb	124 Mb	55 Mb	31 Mb

A single aerial photograph may require as much as 100 megabytes of memory, depending on the pixel resolution. Contrarily, smaller pixels may assure greater accuracy.

After the data are scanned, histograms can be developed with which to adjust radiometric contrast in the formation of a more pleasing overall image tone. Table 14.2 can be of assistance in determining the ground pixel resolution of a standard digital image, and Table 14.3 affirms the expected file size for a black and white digital orthophoto when scanned at various scan sample rates.

14.3.2.3 Ground Control

Ground control is required to rectify (georeference) the imagery to its true geographical position on the earth's surface. Simple rectification (rubber sheeting) is not suitable. Generally, a process known as differential rectification is used. Differential rectification is a phased procedure which uses several XYZ ground control points to georeference an aerial photograph to the earth, thereby creating a truly orthogonal image which can provide accurate measurements throughout its bounds. The exact location and number of ground points required are based upon the scale and accuracy of the final orthophoto as well as the negative scale and number of photo images required to cover the entire project area. Selecting the ground control points is generally not a task for the project manager and should be decided by the photogrammetric technician designing the project based upon his/her experience and equipment.

14.3.2.4 Digital Elevation Model

A suitable DEM must be obtained to provide a vertical datum for an orthophoto. Some projects may allow inclusion of a DEM for the project area that was developed from other imagery. This may be the case when the ground in the project area has not changed significantly between the time the imagery was collected for the DEM and the new orthophoto imagery collection date. However, most large-scale orthophoto projects require a DEM to be developed from the new imagery. This will insure and improve the accuracy of the image rectification.

A DEM for orthophoto rectification does not have to be as dense or as detailed as a terrain model for contour generation. Most projects will only require a coarse grid of points along with breaklines to define areas of abrupt change (i.e., edges of roads, streams, etc.). This task is achieved using the same processes described in Chapter 12, "Photogrammetric Map Compilation." Stereopairs (digital or analog), incorporated into an analytical stereoplotter system or softcopy workstation and rectified to the earth, allow the photogrammetric technician to compile a coarse grid of points and breaklines throughout the project area. The density and spacing of the DEM points and breaklines are dependent largely upon the accuracy requirements, the horizontal scale of the final orthophoto, and the character of the land.

The web site http://www.woolpert.com/news/articles/ar072500.htm provides an insight to combining LIDAR and photogrammetric mapping. A practice application of a project LIDAR and a digital camera for collecting DEM data and creating orthophotos can be found at http:/www.esti.com/library/userconf/proc00/professional/papers/PAP726/p726.htm.

14.3.2.5 Data Merge and Radiometric Correction

The final phase of the orthophoto process is the merger of the digital image and the DEM along with corrections in pixel intensity throughout the image. Software, used to merge the digital raster image with the DEM, makes adjustments in the horizontal location of pixels based upon their proximity to DEM points. This process removes the errors due to displacement and produces an image that is orthogonally accurate. The final step adjusts the intensity of selected groups of pixels in the orthophoto to ensure that seams between mosaicked images are minimized and/or to bring out features of interest or minimize aberrations. Figure 14.3 illustrates a technician creating an orthophoto image electronically.

14.3.2.6 Tiling and Formatting

The conclusive orthophoto image is finally broken into smaller areas that are more convenient to handle by the end user. This process is generally known as tiling or sheeting. Formatting may also be an important task in preparing an orthophoto for submittal to the end user. The final software format and any compression formats should be considered in the design of the orthophoto. Many large orthophoto projects quite often require data to be submitted in the original resolution (i.e., 1-ft pixel resolution) and also resampled and submitted with a 3-ft pixel resolution for quick

Figure 14.3 A photogrammetric technician creating an orthophoto image electronically. (Photo courtesy of authors at Walker and Associates, Fenton, MO.)

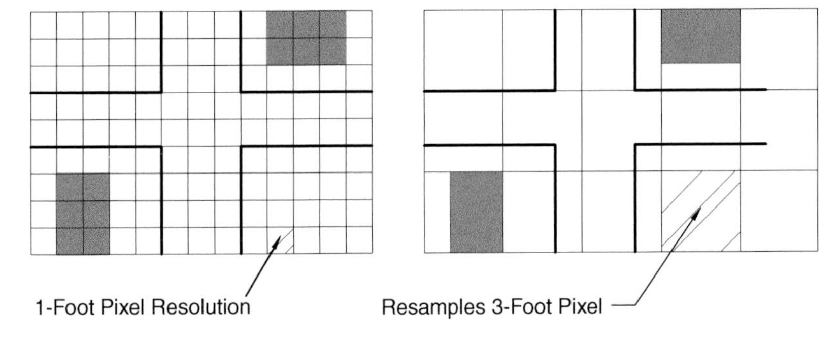

Figure 14.4 An example of an area at 1-ft pixel resolution resampled to 3 ft.

viewing and locating specific areas of concern. Refer to Figure 14.4 for an example of this situation.

14.4 ORTHOPHOTO COST

The cost of orthophoto generation varies widely and is largely dependent upon the scale and accuracy requirements and the availability of source data. Obviously, projects that require the development of new photography and DEM data will be the most labor intensive and costly. However, in many cases suitable orthophotos can be generated from existing imagery and DEM data. The project manager simply needs to ensure that the data will meet the project accuracy and feature needs.

USGS and USACE are two federal agencies that generate both large- and small-scale orthophotos of various parts of the earth. In addition, many federal, state, and local governments collect aerial photography and DEM data that may be used to generate orthophotos. The project manager should investigate these types of sources before investing in new data development when time allows.

CHAPTER 15

Remote Sensing

15.1 REMOTE SENSING

Remote sensing is the science of gathering information from a location that is distant from the data source. Image analysis is the science of interpreting specific criteria from a remotely sensed image. An individual may visually, or with the assistance of computer enhancement, extract information from an image, whether it is furnished in the form of an aerial photograph, a multispectral satellite scene, a radar image, a base of LIDAR data, or a thermal scan.

Remote sensing is a dynamic technical field of endeavor. Between 1995 and 2000 the number of users employed in these combined branches of knowledge rose from 0.7 to 8.1 million, and their commercial application values rose from $3 billion to $12 billion during the same time frame.

The purpose of this chapter is to acquaint the reader with the technology in order to pique his/her interest in pursuing further knowledge, because these procedures may provide sources of pertinent information for the managers and/or technicians involved in mapping or GIS projects. With this in mind, a number of web site references are sprinkled throughout this chapter to start the reader on a voyage of discovery. Most of these web sites are starting points to further guidance.

It should be noted that the remote sensing systems cited in this book do not cover the entire group of remote geospatial data collection systems that is out there waiting to help the project manager in his/her search for applicable digital information. Also, no partiality is intended for those systems and providers that are discussed herein.

This chapter will not dwell on the mechanics of sensors. Rather, it is intended to establish a passing acquaintance with the characteristics of electromagnetic energy, hopefully helpful to the reader in deciding what types of captured information will best fit a particular project needs.

15.2 SEARCHING THE INTERNET

The Internet can be an educational source of pertinent remote sensing information for the project manager to expand his/her technical knowledge.

15.2.1 Tutorials

There are remote sensing tutorials to be found on the Internet, and it may be to the reader's advantage to access a few of these web sites:

- http://hawaii.ivv.nasa.gov/space/hawaii/vfts/oahu/rem_sens_ex/rsex.spectral.1.html
- http://rst.gsfc.nasa.gov/Front/tofc.html
- http://auslig.gov.au/acres/referenc/abou_rs4.htm

The web site http://satellite.rsat.com/rsat/tutorial.html discusses and illustrates spatial analysis, spectral analysis, advanced processing, applications, three-dimensional perspectives, LANDSAT and IRS-1C data fusion, change detection, and various data resolutions.

Refer to web site http://www.ccrs.nrcan.gc.ca for an enlightening tutorial on stereoscopy, radar fundamentals, and stereointerpretation that can be off-loaded for noncommercial instructional purposes.

15.2.2 Applications Dynamics

Remote sensing, along with its entwined sibling sciences (photogrammetry, GPS, GIS), has enjoyed a dynamic upsurge during the past five years. If the reader is interested in technologies related to agriculture, disaster management, environmental monitoring, forestry, mining, transportation, or utilities distribution, he/she would do well to access the web site http://wwwedu.nasa.gov/crsp-wdet/commercial/comApps.html for an extremely instructive session. For each of these technologies this Internet reference provides multipage studies of emerging applications. The Internet web site http://www.flidata.com/ is an enlightening source of information covering the availability of high technology commercial airborne hyperspectral/multispectral imaging systems, advanced remote sensing applications, and airborne data acquisition services applicable to governmental and industrial scientific fields of endeavor:

Agriculture
Defense
Environmental
Forestry
Geographic Information Systems
Mapping
Oceanographics
Research
Terrestrial

For those readers who have a specific interest in forestry applications of remote sensing, log on to http://www.airbornelasermapping.com/Features/ALMSpo01.html for an instructional glimpse at laser mapping.

An instructive Internet reference discussing the principles of remote sensing, with illustrations and images, is found at http://www.sci-ctr.edu.sg/ssc/publication/remote-sense/rms1.html.

15.3 REMOTE SENSING SYSTEMS

Many contemporary mapping technologists collect information with a variety of instrumentation, collectively known as remote sensors. Even though these systems collect digital spatial data in mechanically different ways, all of the captured information is related to the electromagnetic spectrum. Although aerial photos are limited to the 0.4–1.0 μm range, there are other sensors that duplicate this range and still others that can extend their range well into the microwave sector.

The reader may want to access the following key words on the Internet to get a more comprehensive grasp of this subject:

Advanced Very High Resolution Radiometer
Agricultural Research Service Laser Profile
Agricultural Research Service Radiance Transect
Agricultural Research Service Thermal Transect
Airborne Data Acquisition and Registration
Airborne Synthetic Aperture Radar
Airborne Terrestrial Applications Scanner
Airborne Visible/Infrared Imaging Spectrometer
Digital Video Imagery
European Remote Sensing Satellite-1
Japanese Earth Resources Satellite-1
LANDSAT Multisprectral Scanner
LANDSAT Thematic Mapper
MODIS/ASTER Airborne Simulator
Systeme Pour l'Observation de la Terre
Thematic Mapper Simulator/12-Channel Daedelus Multispectral Scanner
Thermal Infrared Multispectral Scanner

These sites can open a lot of doors into the subject of remote sensing methodology and applications.

15.3.1 Thematic Data Collection

A remote sensor is an instrument that gathers thematic information from a distance. Over the years, image analysts have employed various segments of the electromagnetic spectrum to enhance their data gathering capabilities. Commercial use of aerial photography using panchromatic film began about the time of the Civil War, but its extended utilization has come about since the World War I era. World War II saw the beginning of near infrared film and the expanded use of color film. The 1970s began the use of airborne and satellite platforms carrying electromagnetic scanners to collect data from earth. Through the 1980s and 1990s these various spatial vehicles transported scanners utilizing such electromagnetic components as

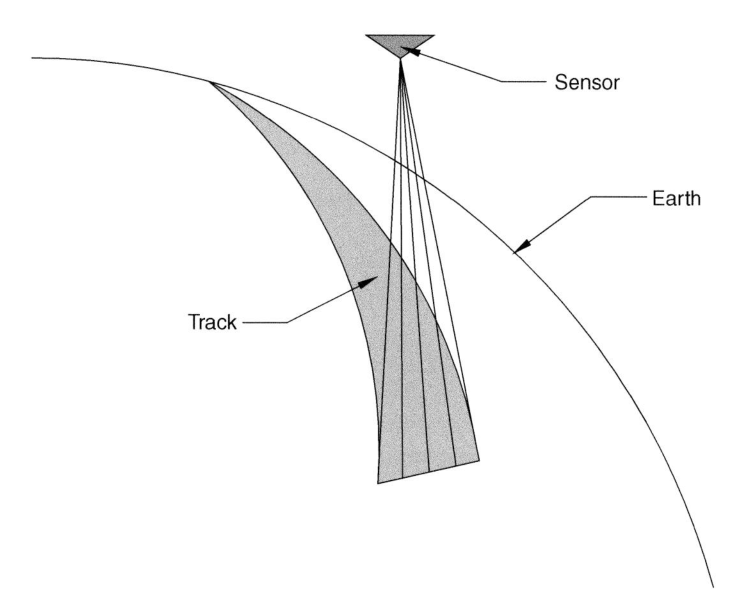

Figure 15.1 A portion of a single scan line.

visible light, near infrared, mid-range infrared, thermal, radar, and LIDAR to collect specialized information.

15.3.2 Scanners

Most scanners operate by catching either radiant rays or return signals and capture information in digital form along scan lines (tracks) forming a continuous orbital path as illustrated in Figure 15.1.

These data are furnished to the buyer as a copy of a segment (scene) of this scanned path information, either in the form of an image or as digital data as shown in Figure 15.2.

Prevailing photogrammetric planimetric and/or topographic mapping is limited to the primary colors of visible light as shown in Table 15.1, but technicians involved in GIS and specialty projects may also want to consider other regions of the electromagnetic spectrum for complementary data sources.

15.3.3 Types of Sensors

Although there are a number of remote sensing systems capable of collecting information, there are two general categories:

- Passive sensors collect natural radiant energy reflected or emitted from a targeted object.
- Active sensors transmit a signal and then receive the reflected response.

There are many different types of data sensors in use, depending upon the purpose of the collected information. Some of the more popular remote sensors currently

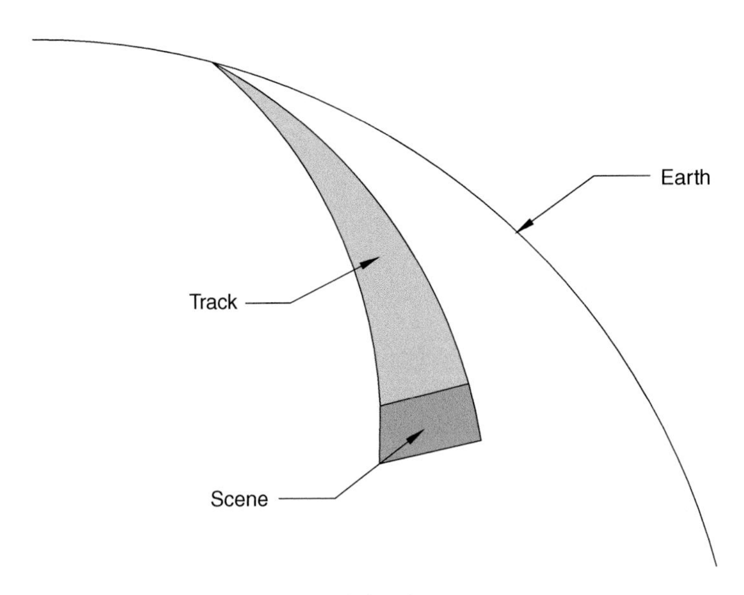

Figure 15.2 A scene from a scanner's orbital path.

Table 15.1 Spectral Bands of Primary Colors

Band	Spectral Range (μm)
Blue	0.4–0.5
Green	0.5–0.6
Red	0.6–0.7

employed in data capture are listed herein, but it is not intended that all available systems be included. Those shown represent a sampling of the different segments of the electromagnetic spectrum that are used in analyzing ground characteristics.

15.3.3.1 Aerial Camera

An aerial camera is a passive sensor that collects a direct, continuous tone pictorial image in the visible light (0.4–0.7 μm) range. Through the use of proper film, the camera can also create a photographic near infrared image composed of visible green (0.5–0.6 μm), visible red (0.6–0.7 μm), and near infrared (0.7–1.0 μm) light.

15.3.3.2 Video Camera

A video camera can be installed in an aircraft. This passive sensor records a continuous swath of raster data covering a moving scene of the terrain, and the videotape can be played on a graphic screen much like a video movie. Digital video systems today are often used to collect, manipulate, and analyze data in the black and white, natural color, and color infrared ranges of the spectrum. Since the video image is a raster file, it can be segmented and imported into a CADD/CAM/CAD environment.

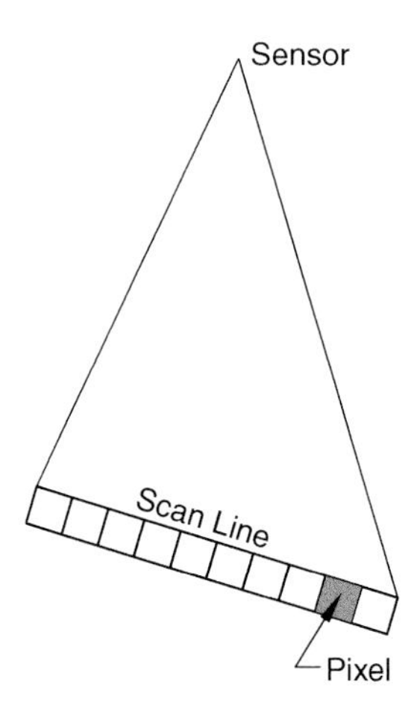

Figure 15.3 Scanned pixels.

15.3.3.3 Scanners

Scanners are passive sensors that capture the reflected or emitted energy intensity from observed objects into digital picture elements called pixels. Scanner data can be viewed as a pictorial rendition on a computer screen or generated as a hardcopy counterpart. Data gathered as groups of pixels are termed raster data. Figure 15.3 is a schematic representation of a scanned data line with a single pixel blackened.

Thematic Mapper/Multispectral Scanner

Thematic mappers (TM) and multispectral scanners (MSS) are passive scanning systems that collect raster data in several selected bandwidths simultaneously between visible light and thermal bandwidths (0.4–8.0 μm). Refer to Figure 15.4 for a schematic of a rudimentary MSS. These sensors have been deployed on several systems of earth resource satellites.

Operational scanning systems are considerably more complex than this simplistic diagram implies. A revolving mirror makes successive raster sweeps of the terrain as the carrier moves forward. Pixels of reflected and/or emitted energy wave bundles pass through the system aperture to be reflected off the surface of the rotating mirror onto a beam-splitting mirror that reflects specific wavelengths and transmits others. This grate deflects the visible portion of the spectrum, while the thermal passes on to be collected by thermal detectors. The visible waves pass through a prism where they are separated into various colors, which are collected by visible light detectors. Data are stored in groups of waveband ranges.

Thermal Scanner

A thermal scanner is a passive scanner that collects raster data in the longer infrared wavelengths (8–13 μm range) which are actual temperature radiations emitted

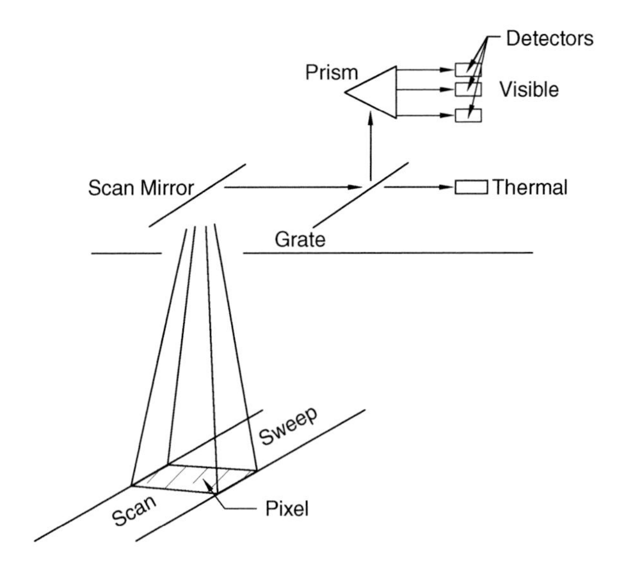

Figure 15.4 Rudimentary components of an MSS.

from an object. Since this scanner senses heat emissions, it can be employed during daylight or darkness.

Radar

The web site http://southport.jpl.nasa.gov/drsc/imagingradarv3.html presents a basic dissertation on the subject of imaging radar. Synthetic aperture radar (SAR) is an active scanner that transmits and receives its own signals in several bands within the microwave range (1 mm to 1 m range). The receiver records a continuous swath of raster data covering a moving scene, and the data tape can be played on a graphic screen much like a video movie. Radar scans can be segmented and imported into a CADD/CAM/CAD environment.

This system is capable of piercing clouds and penetrating to certain depths in the soil mantel and can be operated during daylight or darkness, which partially accounts for its increasingly wider usage. Reflected radar signals are measurable, thus enabling mappers to calculate geographic coordinate values of ground features.

Integrating radar sweep data into a DTM structure can generate interpolated contours covering designated tracts.

Radar has a number of capabilities, which makes it a valuable sensor in a variety of applications, some of which are listed in Table 15.2. The web site http://www.sandia.gov/radar/sarapps.html offers more information.

Light Detection and Ranging

LIDAR (light detection and ranging*), a relatively recent innovative technique for the collection of digital elevation data, shows great promise for terrain mapping

* Log on to the Internet with this key search phrase to open various references pertaining to LIDAR.

Table 15.2 General Applications of SAR

Military	Reconnaissance
	Surveillance
	Targeting
	Buried arms caches and mines
	Underground bunkers
Treaty verification	Weapons nonproliferation
Guidance	All-weather navigation
Penetration	Foliage
	Soil
	Underground utilities
Environmental	Crop characteristics
	Deforestation
	Ice flows
	Oil spills
	Oil seepage

applications. LIDAR is capable of producing a mass of spatial points that may be used as basic elevation data for production of surface models such as DEMs, DTMs, and computer software-generated contours. Additional ancillary products may be developed from LIDAR elevation data sets with the use of specific software techniques. The light energy that is emitted by the laser strikes a terrain surface. A portion of the energy is absorbed by the surface. The amount of absorption is partially dependent upon the type of surface that the light strikes. The remaining energy reflects off the terrain surface and is captured by the sensor. Intensity images are software-produced images created by assigning colors or shades of gray to the amount of energy returned to the sensor from laser light pulses. Intensity images can provide a crude pictorial of the earth surface that may have use as a reconnaissance or planning tool. The project manager should remember that LIDAR data are simply elevation data used in the preparation of elevation products. Raw LIDAR data is not an end product itself, and in fact in most cases is of little use in photogrammetric mapping. It is important for project managers to understand that LIDAR is only one of several posible tools that can be used to collect elevation data. The web site http://lidar.woolpert.com discusses benefits, system components, specifications, accuracy, flight layout, post processing, and quality control as they pertain to LIDAR operations.

The airborne LIDAR system is composed of multiple interfaced systems, which may consist of the following:

1. An infrared laser discharging a stream of focused pulses at a rotating mirror, which scatters them across a swath on the ground. When the receiver unit recaptures the reflected rays, a discriminator and a time interval meter measure the elapsed time between the transmitted signal and the reflected echo.
2. As the flight is in progress an inertial reference system (IRS) automatically maintains a constant record of the pitch, roll, and heading of the aircraft.
3. Throughout the flight, a kinematic ABGPS locks onto at least four navigation satellites, thereby constantly documenting the spatial position of the aircraft.

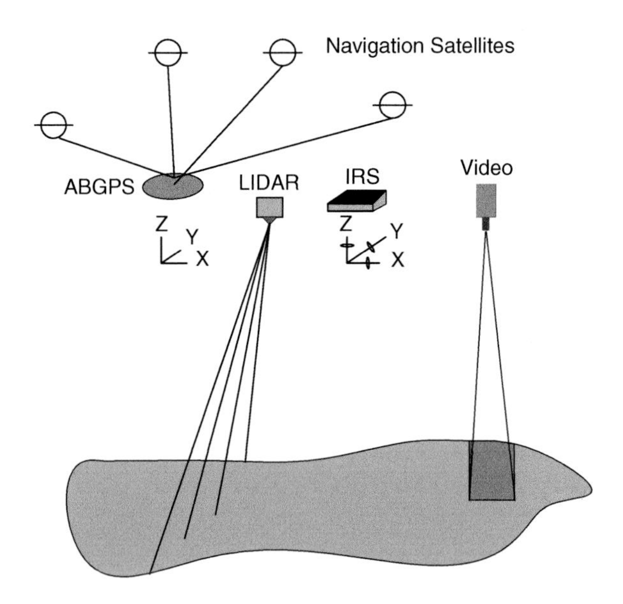

Figure 15.5 Schematic of the components of an airborne LIDAR system.

4. An imagery collection system (analog camera, digital camera system or color video camera) records the terrain along the track of the LIDAR scan. Imagery may be required for quality control of final processing of the data and planimetric feature collection. Many projects may not require the collection of this type data or may be able to make use of existing imagery.

Figure 15.5 illustrates the concept of how the separate airborne components are linked in an operational mode.

During the mission one or more ground GPS stations are linked into the system to assure dependable referencing of the airborne package to the earth. Once the flight data are recorded, appropriate software manipulates the combined data and creates a spatial coordinate at each ray point. The accumulated digital points are stored in a massive database of ground stations. LIDAR projects that are designed properly can economically generate digital terrain models with vertical accuracies as close as 6 in. or less, and horizontal accuracies within 1/1000 of the flight height.

LIDAR offers some advantages over aerial photography in creating topographic maps:

- Overflights can be scheduled at almost any time because LIDAR is uninhibited by the time of day, sun angle, certain types of vegetation, or less than ideal weather conditions. Rain, and flights that require an altitude in or above cloud cover, are unsuitable conditions for LIDAR collection.
- Foliage penetration is possible. Penetration of foliage may vary depending upon many factors, including the specifications for the laser to be used, type of foliage, swath width, and flight height. Penetration of foliage is one of the most important features of a LIDAR terrain data collection project. Many LIDAR projects have as final

products a model of the earth's surface without trees, foliage, and planimetric features. The LIDAR industry is constantly improving collection and processing systems for penetration of foliage and the accurate removal of trees and planimetric features.
- Terrain data may be collected and processed in an expeditious manner, thus possibly reducing project completion time.
- Flights are not inhibited by restricted right of entry or remote sites.

Since the rays are measured to a point where they are reflected by an object — foliage, structures, ground — editing of the LIDAR data must be done in a highly adept manner. Contractor LIDAR data processing techniques are often unique. The speed and accuracy of these techniques are critical to the success of a LIDAR elevation data collection and final product generation process.

Forward Looking Infrared

Forward looking infrared (FLIR) is a passive scanner which converts incident thermal (heat) rays into real-time video signals. This system may be brought into play during daylight or darkness and can utilize airplanes, helicopters, or ground vehicles as carriers employed in police work, search and rescue missions, wild game census, and environmental studies where differential shades of temperature values segregate primary interest objects from a cluttered background. Consult the web site http://www.flir.com/resources/InfraredEverywhere.htm for further information about this system.

15.4 AERIAL PHOTO IMAGE SCANNING

Although there is a growing popularity for digital cameras in aerial photography projects, most aerial photography is accomplished with an analytical camera. Spurred by the transition toward softcopy systems, there is a growing trend to scan aerial photographs with which to superimpose raster images on vector mapping information either on the computer monitor or hardcopy data plots. Photogrammetric mapping projects require high-resolution scanning with generally between 800 and 1600 points per inch. Scanning at these resolutions requires large quantities of hard disk storage. Photogrammetric mapping projects typically require a significant number of photographic images. Photogrammetric workstations that require the incorporation of scanned imagery will necessarily mandate significant processing and digital data storage capacity.

A series of simple formulae calculate the amount of disk storage for photo scanning. Equation 15.1 determines the number of raster points per line, based upon pixel resolution.

$$p_l = p_i * L \tag{15.1}$$

where:
p_l = points per line
p_i = resolution (points per inch)
L = length of image (inches) in line of flight

The number of scan lines is reckoned with Equation 15.2, dependent upon raster resolution and width of the image frame.

$$d_l = p_i * W \tag{15.2}$$

where:
d_l = lines of data
p_i = resolution (points per inch)
W = width of image (inches) perpendicular to flight

The total bytes of scanned data are calculated with Equation 15.3, based upon points per line and number of lines.

$$d_b = p_l * d_l \tag{15.3}$$

where:
d_b = total bytes of scanned data
p_l = points per line
d_l = number of lines of data

As an example, a total panchromatic aerial photo is to be scanned at a resolution of 200 points per inch. A photo encompasses a 9-in. square area.

$p_l = p_i * L = 200 \times 9 = 1800$ points
$d_l = p_i * W = 200 \times 9 = 1800$ lines
$d_b = p_l * d_l = 1800 \times 1800 = 3{,}240{,}000$ bytes

Each point requires 1 byte of storage, so disk space for this single panchromatic photo requires 3.24 megabytes at a resolution of 200 points per inch.

It is not always necessary to scan the entire photo, which would lessen storage requirements, but sometimes multiple photos are necessarily scanned to form mosaics. This technique adds to the storage. Color photos require a separate scan for each primary color of red, green, and blue. Therefore, color storage requires three times as much disk space as panchromatic.

15.5 SATELLITE IMAGERY*

Every portion of the surface on the earth receives solar radiation that is reflected or absorbed and emitted in specific wavelengths, some of which are invisible to the human eye. A surface's characteristic spectral signature is made up of those specific wavelengths that can be recorded by satellite-mounted sensors. The recorded spectral signatures are subsequently processed into photograph-like images.

* Log on to the Internet key phrase "satellite remote sensing systems" which opens numerous informative web sites on this subject.

15.5.1 Data Format

Satellite images are collected in raster format, which is a matrix of thousands of individual picture elements called pixels. The ground area covered by each pixel determines the resolution of the pixel. For instance, if the image resolution is 30 m, each picture element is restricted to an area on the ground covering 30 m, about 100 ft, square.

Each pixel contains a single unit of information which represents the dominant spectral signature for the corresponding area on the earth's surface. The information content of an individual pixel is in digital form, usually 8-bit, so that it can be analyzed by a computer or converted to photographs for visual analysis. The information set captured by a pixel is composed of its XY position and the brightness value.

15.5.2 Spectral Bands

Satellites collect data in groups of spectral bands. In a natural color image there would be three bands of data (red, green, and blue), each showing various intensities of the pertinent color. By the same token, a false color photographic image would also contain three bands of data (near infrared, red, and green), each showing various radiances for the pertinent layer.

15.5.3 Georeferencing Satellite Data

Satellite data are subject to a number of errors, but vendors process the raw data through standard algorithms and cleanse the information prior to delivery. Satellite images can be georeferenced to earth coordinates. Once the image is georeferenced to the ground, these causal errors are normally not significant at the mapping scales involved. In many situations this can be a "rubbersheet" scaling process rather than a true displacement rectification, whereby the image is best fit to several ground control points. In other situations, some processes that use orthophoto generation of the image produce a truly scale rectified image.

15.5.3.1 Advantages of Satellite Scenes

Presuming that the inherent accuracy of satellite data conforms to the mapping specification demands, satellite data can be a valuable tool in many ways.

15.5.3.2 Pictorial Image

Satellite raster imagery provides a pictorial simulation that can be overlaid on GIS/LIS vector themes so that the viewer can see the image and line drawing simultaneously. Hardcopy plots of the image data can be created.

15.5.3.3 Change Detection

Historical satellite imagery for different passes over the same site is available for purchase. Change detection information can be gleaned from these time-lapse scenes.

15.5.3.4 Perspective Views

Three-dimensional perspective maps can be generated, looking at the data scene from any angle, with a spatial image or a GIS file draped over it.

15.5.3.5 Screen Digitizing

After an image is registered on the screen data can be digitized from it, thus creating polygonal thematic layers.

15.5.4 Restrictions

Two restrictions must be realized when using satellite imagery:

- These operations require tremendous amounts of RAM and hard disk storage.
- Data compatibility is only as good as the most inaccurate information.

15.6 SATELLITE SYSTEMS

Currently, a number of major earth resource satellite systems revolve around the world in sun-synchronous paths. It should be noted that those discussed herein are not the total collection of orbiting satellites, but are intended to inform the reader about the different types of electromagnetic information that is available to the public.

15.6.1 LANDSAT

For further information about the LANDSAT satellite umbrella, refer to:

http://www.friends-partners.org/~jgreen/landsat.html
http://geo.arc.nasa.gov/sge/landsat/lpsum.html
http://www.fes.uwaterloo.ca/crs/geog376/EOSatellites/Landsat.html

In 1972 the United States launched LANDSAT 1, an optical satellite, the first of what would become a series of earth resources satellites. Since then six other satellites in this series were put into service, so at times multiple sensing systems were in orbit simultaneously. Figure 15.6 indicates the launch intervals of LANDSAT vehicles 1–7 and follows their operational longevity. LANDSAT 8 is scheduled for flight in about 2004.

These satellites fly in generally a north/south orbit over the sunlit portion of the earth while collecting electromagnetic data along a scan line. Data from the LANDSAT vehicles are transferred to earth stations for processing and distribution.

Table 15.3 indicates the payloads carried by each of the satellites in the LANDSAT series. The data collection instruments referenced in the table are:

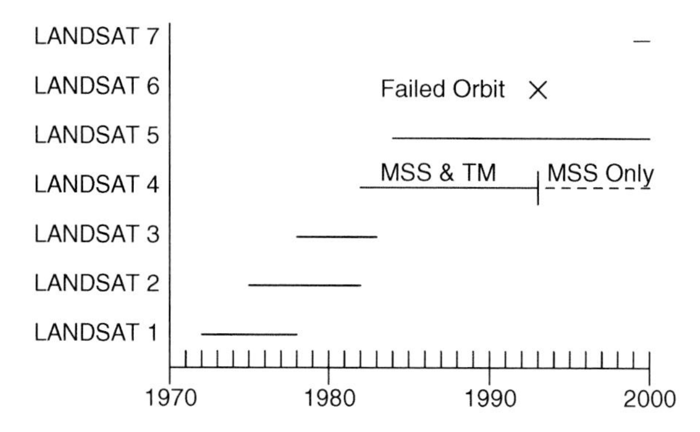

Figure 15.6 Launch years and longevity of the LANDSAT series.

Table 15.3 Sensor Payload Aboard Each of the Vehicles in the LANDSAT Series

System	RBV	MSS	TM	ETM	ETM+
LANDSAT 1	√	√			
LANDSAT 2	√	√			
LANDSAT 3	√	√			
LANDSAT 4		√	√		
LANDSAT 5		√	√		
LANDSAT 6				√	
LANDSAT 7					√

RBV (Return Beam Vidicon)
MSS (Multispectral Scanner)
TM (Thematic Mapper)
ETM (Enhanced Thematic Mapper)
ETM+ (Enhanced Thematic Mapper Plus)

Table 15.4 cites the resolution of the sensory systems of the various LANDSAT series. The resolution discriminates between data collected in panchromatic (pan) and multispectral (ms). LANDSAT 1 through LANDSAT 5 provide data in both panchromatic and multispectral mode at the same resolution.

The spectral characteristics of each vehicle in the LANDSAT constellation can be found at the web site http://www.ccrs/nrcan.gc.ca/ccrs/tekrd/satsens/landsate.html.

The orbital altitude and revisit* schedule of the LANDSAT series can be found in Table 15.5.

Some applications that have been used or might be considered for use with LANDSAT imagery are listed in Table 15.6.

A USGS guide to LANDSAT coverage can be found by referencing "Thematic Mapper LANDSAT Data: Search Criteria" at http://edcwww.cr.usgs.gov/Webglis/glis-bin/search.plLANDSAT_TM on the Internet.

* Revisit is the time interval between successive data captures of the same scene.

Table 15.4 Resolution of Instrument Packages in the LANDSAT Series

System	Instrument	Resolution (meters)
LANDSAT 1	RBV	80
	MSS	80
LANDSAT 2	RBV	80
	MSS	80
LANDSAT 3	RBV	30
	MSS	80
LANDSAT 4	MSS	80
	TM	30
LANDSAT 5	MSS	80
	TM	30
LANDSAT 6	ETM	15 (pan)
	ETM	30 (ms)
LANDSAT 7	ETM+	15 (pan)
	ETM+	30 (ms)

Table 15.5 Altitude Above the Earth's Surface and Revisit Interval of Each of the LANDSAT Series

Series	Altitude	Revisit
LANDSAT 1	917 km	18 days
LANDSAT 2	917 km	18 days
LANDSAT 3	917 km	18 days
LANDSAT 4	705 km	16 days
LANDSAT 5	705 km	16 days
LANDSAT 6	Inoperable	Inoperable
LANDSAT 7	705 km	16 days

15.6.2 SPOT

The French-based SPOT (Systeme Pour l'Observation de la Terre) has launched a constellation of four optical earth resources satellites into orbit. Refer to the SPOT web pages http://www.spot.com, http://edcwww.cr.usgs.gov/glis/hyper/guide/spot or http://version0.neonet.nl/ceos-idn/sources/SPOT_3.html for information about this earth observation system. Refer to Figure 15.7 for the launch and decommission dates of the SPOT series.

The payload on each satellite consists of two high-resolution sensor systems. Working independently of one another, these systems add a high degree of flexibility for customized three-dimensional data collection. Characteristics of the SPOT series can be located by logging on to the web site http://www.ccrs.nrcan.gc.ca/ccrs/tekrd/satsens/sats/spote.html.

15.6.2.1 Off-Nadir Viewing

Initially, this series of data collection satellite systems gathered continuous monoscopic raster vertical image information. Once the SPOT satellite was launched,

Table 15.6 Some Applications that Have Been Used or Might Be Considered for Use with LANDSAT Imagery

Hydrology	Agriculture
Watershed modeling	Crop assessment
Wetland conditions	Crop location
Snow pack conditions	Crop damage
Lake ice	Yield estimates
River ice	Temporal change
Sea ice	Compliance Monitoring
	Farming activity
Forestry	Soil Condition Monitoring
Mapping	Tillage practice
Forest inventory	
Forest cover typing	Disaster Management
Clearing location	Flood mapping
Pathogen location	Extent
	Damage
Engineering	Oil spill monitoring
Route location	Detection
Pipelines	Mapping
Power lines	Forest fires
Roads	Burn delineation
Utilities	Damage assessment
Geology	Land Use/Land Cover
Mapping	Land use monitoring
Structure mapping	Use/cover patterns
Surfacial bedrock mapping	Temporal change
Lineament identification	Land cover delineation
Landform delineation	Vegetation
Surfacial material	Cover types
Geological Hazard	Base mapping
Landslide hazard	Land use
Coastal erosion	Land cover
	Cultural features

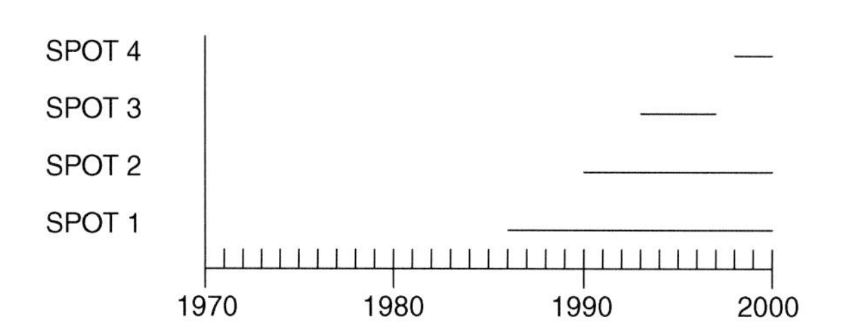

Figure 15.7 Launch years and longevity of the SPOT series.

this capability changed. Aside from amassing vertical imagery, this system can also be directed to collect side-looking information in an area covered by a previous pass.

Data collected by SPOT satellites, when in normal operational mode, are monoscopic. Unlike LANDSAT (nadir viewing only), SPOT can view both nadir (vertical) and off-nadir (oblique). Upon specific request this allows the sensor system to capture the same scene on several different passes, with the number varying from seven on the equator to eleven at 45° latitude, which allows for stereoscopic coverage. Presented with this stereoscopic faculty, mappers can compile planimetric and/or topographic maps from spatial platforms soaring 400–500 mi above the ground. Granted, this ability is limited to relatively small-scale mapping, but there are still inaccessible areas in the world that demand a use for this potential.

15.6.2.2 Applications

As noted above, Table 15.6 lists some specific applications to which LANDSAT imagery has been used or might be considered applicable. By the same token, Table 15.7 lists some specific applications to which SPOT imagery has been used or might be considered applicable. Some other applications may be in the general fields of cadastral mapping, land cover mapping, telecommunications, surveillance, natural hazard assessment, or many others requiring a view from above. A SPOT

Table 15.7 Some Applications that Have Been Used or Might Be Considered for Use with SPOT Imagery

Technology	Application
Agriculture	Crop forecasting
	Productivity monitoring
	Soil moisture assessments
	Crop damage assessments
Cartography	Topographic mapping
	Terrain simulations
	Terrain modeling
	Thematic mapping
	Map updating
Forestry	Harvest logistics
	Stand density
	Yield estimate verification
	Disease assessment
	Fire damage assessment
Geology	Oil and gas exploration
	Structural mapping
	Engineering studies
	Hazards analysis
	Tectonic studies
Urban planning	Land use mapping
	Impervious surface modeling
	Siting studies
	Demographic change detection
	Cultural change detection
Water and environment	Wetlands mapping
	Habitat mapping
	Pollution monitoring
	Resource assessment
	Hydrological studies
	Coastal studies

scene can be transformed into a three-dimensional view if integrated with corresponding DEM information.

SPOT 5, scheduled for a 2002 launch, will offer several innovative features:

- 60 × 60 km imagery scenes
- 2.5-, 5-, 10-, and 20-m resolution
- Daily global coverage at 1-km resolution
- Worldwide DTMs

SPOT's panchromatic 15-m resolution permits the imagery to be used for the production of thematic mapping with detail location accuracy comparable to map scales of cartographic work at 1:100,000 scale and map updating at 1:50,000 scale. SPOT's side-looking capability permits stereoscopic imagery and allows three-dimensional viewing and interpretation of terrain and cultural features from any location on the earth. This stereoscopic imagery is being used to produce topographic maps with contour intervals as low as 10 and 20 meters. It is also used in digital terrain modeling and the production of three-dimensional perspective views for terrain simulation, strategic planning, and impact assessments. When there is an interest in using only specific areas of satellite scenes, relevant areas can be cut out of the total scene.

15.6.2.3 National Oceanic and Atmospheric Agency

Log on to the web site http://publicaffairs.noaa.gov for greater in-depth information about the National Oceanic and Atmospheric Agency (NOAA) sensing system. This meteorological satellite system launched by the NOAA carries a pair of advanced very high resolution radiometers (AVHRR) with resolutions of 1.1 and 4.0 km. Sensitivity ranges are visible, near infrared, and three thermal bands. Refer to the web site http://www.ccrs.nrcan.gc.ca/ccrs/tekrd/satsens/sats/noaae.html for the visible, near infrared, and infrared characteristics of the NOAA series.

The NOAA sensing system revisits the same scene twice daily and produces a continuous east/west swath 1490 mi (2400 km) wide from an altitude of 900 mi (1450 km) above the earth. Imagery from these satellites is compatible only with ultra small-scale mapping. To get information concerning the systems or to order data, contact:

NOAA, Satellite Data Services Division
5627 Allentown Road
Camp Springs, MD 20746

15.6.2.4 Indian Remote Sensing

Having had encouraging success in testing demonstration resource satellites, Bhaskara 1 in 1979 and Bhaskara 2 in 1981, the India Department of Space set into motion the Indian Remote Sensing (IRS) Satellite program with the implementation of the National Natural Resources Management System. With the goal of boosting the national economy, this remote sensing system was designed to furnish informative

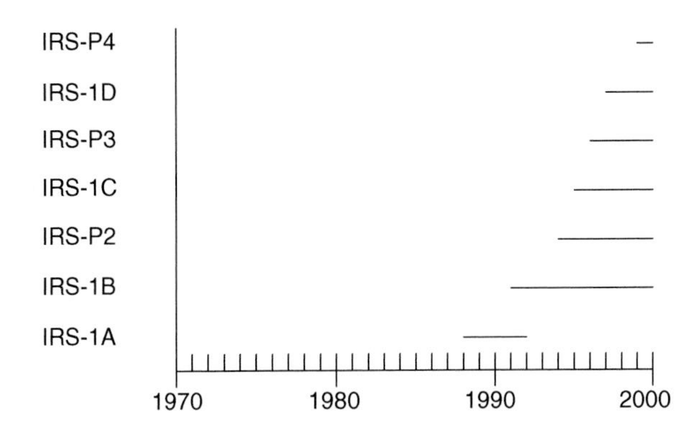

Figure 15.8 Launch years and longevity dates of the IRS series.

Table 15.8 Payloads of Those IRS Remote Sensor Satellites Operating at the Beginning of 2000

	LISS-I	LISS-II	LISS-III	PAN	WIFS	MOS	OCM	MSMR
IRS-1B	●	●						
IRS-P2		●						
IRS-1C			●	●	●			
IRS-P3					●	●		
IRS-1D			●	●	●			
IRS-P4							●	●

electromagnetic data to scientists in such ventures as agriculture, forestry, ecology, geology, watersheds, marine fisheries, and coastal management.

Since 1988 the Indian Space Research Organization has launched a constellation of seven satellites, with two more planned by 2004. Figure 15.8 indicates the launch years and longevity of the IRS series. The IRS satellites carry three instruments in four bands, covering visible and near infrared. One device on the early series covers a ground swath 93 mi (148 km) wide with a resolution of 72.5 m, and the other two devices cover parallel swaths 46 mi (74 km) wide with a 1.5-km overlap at a resolution of 36.25 m. The latest models (IRS-1C and IRS 1-D) produce panchromatic image data at less than 6-m resolution and multispectral image data ranging from 25 to 200 m. Table 15.8 notes the payloads of those IRS remote sensor satellites operating at the beginning of 2000. The system acronyms are:

- LISS — Linear imaging self-scanning
- PAN — Panchromatic
- WIFS — Wide field sensor
- MOS — Modular optoelectronic sensor
- OCM — Ocean color monitor
- MSMR — Multifrequency scanning microwave radiometer

With the subsequent successful operation of this series of satellites, the worth of their captured information has blossomed into numerous rational applications, such as:

- Crop acreage measurement
- Crop yield estimation
- Agro-climatic planning
- Drought warning and assessment
- Flood control, risk zones, and damage assessment
- Watershed management
- Water resources management
- Prediction of snowmelt runoff
- Irrigation management
- Wetland mapping
- Land use/land cover mapping
- Wasteland management
- Fisheries management
- Mineral prospecting
- Forest resource surveys
- Urban planning
- Environmental impact

IRS-1B

Refer to http://csre.iitb.ernet.in/isro/irs-1b.html on the Internet for the characteristics of IRS-1B sensors.

IRS-P2

Refer to http://csre.iitb.ernet.in/isro/irs-p2.html on the Internet for the characteristics of IRS-P2 sensors.

IRS-1C

Refer to http://csre.iitb.ernet.in/isro/irs-1c.html on the Internet for the characteristics of IRS-1C sensors.

IRS-P3

Refer to http://csre.iitb.ernet.in/isro/irs-p3.html on the Internet for the characteristics of IRS-P3 sensors.

IRS-1D

Refer to http://csre.iitb.ernet.in/isro/irs-1d.html on the Internet for the characteristics of IRS-1D sensors.

IRS-P4

Oceansat-1 measures physical and biological ocean criteria on eight spectral bands on the OCM. Refer to http://csre.iitb.ernet.in/isro/irs-p4.html on the Internet for the characteristics of IRS-P4 sensors.

Future

The slated launch of IRS-P5 (Cartosat-1) and IRS-P6 (ResourseSat) in the 2000–2002 time frame adds cadastral level capabilities for cartographic mapping and crop/vegetation analysis.

15.6.2.5 Earth Remote Sensing

Within the umbrella of the European Space Agency a group of nations has put into operation a pair of Earth Remote Sensing (ERS) satellites. The first, ERS-1, was launched in 1991 and went defunct in 2000. ERS-2, very similar to ERS-1, was launched in 1995 and is still operable. An interested reader may wish to refer to the web site http://earthnet.esrin.esa.it/eeo4.10074 for more information.

The payload in this vehicle includes several electronic instruments to carry out a number of functions:

1. A C-band (56-mm wavelength) SAR with a spatial resolution of 30 m
2. A radiometer (visible and infrared) that records sea surface temperatures and vegetative land cover
3. A wind scatterometer
4. A radar altimeter that measures wave magnitude
5. An absorption spectrometer that detects upper level ozone, trace gases, and aerosols
6. A microwave sounder that detects atmospheric humidity
7. A range device that ascertains orbit and trajectory information
8. A laser reflector that pinpoints satellite position

The web site http://www.ccrs.nrcan.gc.ca/ccrs/tekrd/satsens/sats/erse.html presents the characteristics of the ERS series.

It was intended that these sensors would provide information about the oceans, iceflows, and land resources. This active electronic system is capable of collecting reliable resource data pertinent to the technologies involving meteorology, geology, vegetation, hydrology, land use, oceanography, and glaciology regardless of time of day or the presence of cloud cover, haze, or smoke.

To get information, archived back to 1992, concerning the systems or to order data, contact Space Imaging EOSAT (Section 10-13) at:

Space Imaging EOSAT
12076 Grant Street
Thornton, CO 80241

15.6.2.6 IKONOS-2

Originally planned by Space Imaging EOSAT as a dual vehicle system, IKONOS is now composed of a single functioning satellite. Immediately subsequent to its vernal launch in 1999, telemetry from the rocket carrying IKONOS-1 ceased and was never regained. Due to technical difficulties the rocket plunged into the Pacific Ocean shortly after liftoff. IKONOS-2 was launched six months later and has continued to operate successfully since then.

Table 15.9 Width and Resolution of Spectral Bands of IKONOS-2 Digital Camera

Band	Spectral (μm)	Resolution (m)
Monochromatic	0.45–0.90	1
Multispectral		4
Blue	0.45–0.52	4
Green	0.52–0.60	4
Red	0.63–0.69	4
Near infrared	0.76–0.90	4

A visit to the Internet under the keywords "IKONOS 1 satellite," "IKONOS 2 satellite," or "Space Imaging EOSAT" would add to the reader's knowledge.

The payload's major component is, in essence, a very high resolution digital camera which is capable of gathering gray-scale and multispectral digital data from an altitude of 422 mi above the earth. The data collector is a push-broom electro-optical camera with a 10-m focal length folded by mirrors into a 2-m package. It sweeps a 700-km swath and is capable of pivoting to collect cross-track data.

Applications of this system would be similar to any those of LANDSAT or SPOT, but its very high resolution gives it a distinct advantage in analyzing information in GIS and other scientific resource projects. Table 15.9 notes the path width and resolution of spectral bands of IKONOS-2 digital camera.

Refer to http://www.erdas.com/news/Ikonos_image.html on the internet to view a striking IKONOS-2 monochromatic image.

To get information concerning the systems or to order data, contact Space Imaging EOSAT.

15.6.2.7 RADARSAT

RADARSAT's payload, a Canadian optical satellite system carried in the RADARSAT-1 vehicle, utilizes a C-band (5.6 cm) SAR, an active sensor which discharges microwave signals to the ground and captures return signals. These signals, from a single wavelength frequency, can be translated into a panchromatic image that may then be colorized by integration with data from other sources such as SPOT or LANDSAT. This system's long wavelength allows its use during the day or night under most atmospheric conditions. Figure 15.9 is a RADARSAT image scene in a Fairbanks, AK mining district.

Some of the applications that have been used or might be considered for use with RADARSAT imagery can be found in Table 15.10. Those interested in learning more about these applications may wish to log on to http://www.rsi.ca/classroom/class.htm for a more in-depth study of these applications.

Data, which is processed in near real time, is furnished in six grades of resolution ranging from fine (8 m) to coarse (100 m) in scene sizes from 50 × 50 km to 500 × 500 km. Archived data dates back to 1996.

RADARSAT-2, which will allow capabilities of 3-m resolution, is scheduled for launch in 2002.

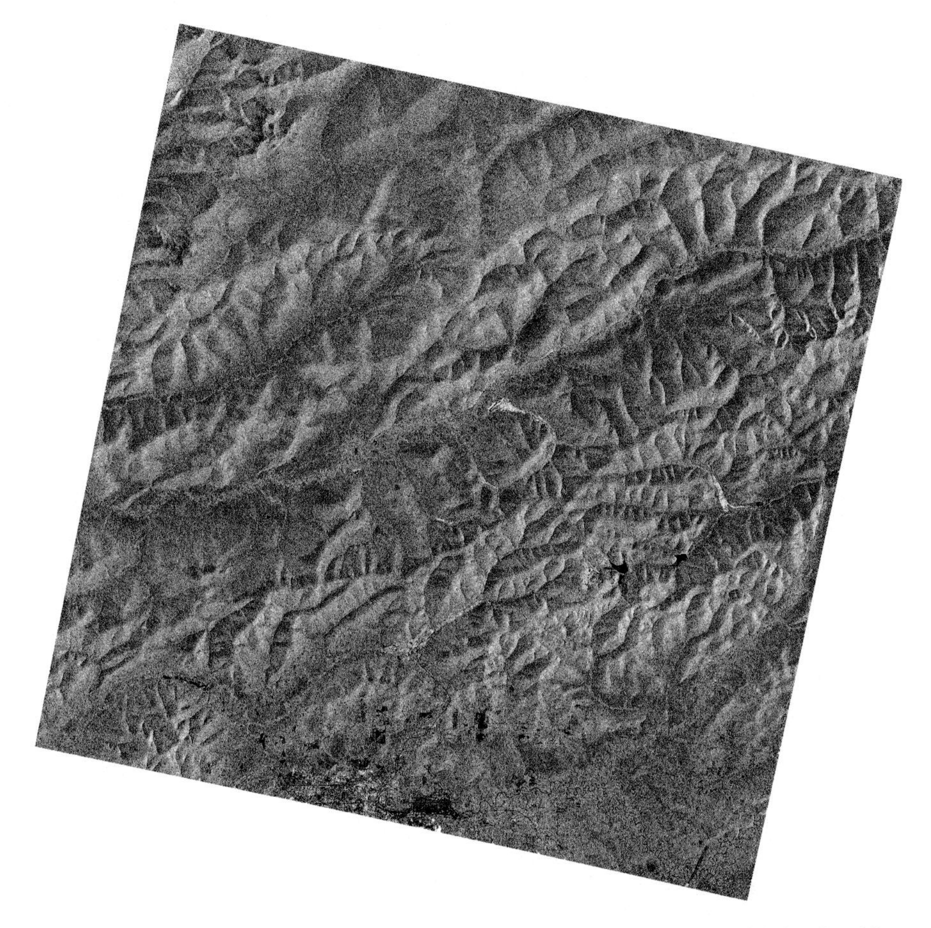

Figure 15.9 RADARSAT image scene in a Fairbanks, AK mining district provided to AeroMap U.S. by Space Imaging EOSAT. (Courtesy of AeroMap U.S., Anchorage, AK.)

Besides RADARSAT imagery, this organization also offers products from LANDSAT, SPOT, ERS, and JERS. Headquarters and client services offices can be contacted at:

RADARSAT International
13800 Commerce Parkway
MacDonald Dettwiler Building
Richmond, British Columbia
V6V 2J3, Canada

Headquarters	Client Services
Tel: (613) 231-500	(604) 244-0400
Fax: (604) 231-4900	(604) 244-0404
E-mail: info@rsi.ca	info@rsi.ca

Table 15.10 RADARSAT Applications

Hydrology	Agriculture	Geology
Watershed modeling	Crop assessment	Geological mapping
Soil moisture	Crop type	Structure mapping
Wetland conditions	Crop damage	Surfacial bedrock maps
Snow pack conditions	Land use monitoring	Lineament identification
Fresh water ice	Temporal change	Hydrocarbon exploration
Lake ice	Compliance monitoring	Sedimentology maps
River ice	Farming activity	Mineral exploration
	Land use elevation	Quaternary mapping
Forestry	Soil condition monitoring	Landform delineation
Recon mapping	Tillage practice	Surfacial material
Terrain analysis	Soil moisture	Geological hazard
Forest cover types		Seismic zones
Commercial forestry	Disaster Management	Landslide hazard
Clearing mapping	Flood mapping	Coastal erosion
	Extent	
Marine	Damage	Land Use/Land Cover
Coastal zone monitoring	Oil spill monitoring	Land use monitoring
Vegetation mapping	Detection	Use/cover patterns
Mapping	Mapping	Temporal change
Ship target detection	Emergency response	Land cover delineation
Vessel location	Forest fires	Vegetation
Wake detection	Burn delineation	Cover types
Aquaculture detection	Damage assessment	Base mapping
Location		Land use
Mapping		Land cover
Ocean circulation		Cultural features
Feature ID		

An interested reader may wish to contact the Internet page "How to order RADARSAT-1 Data" at http://www.space.gc.ca.sectors/earth_environment/radarsat/order_d_ata/default.asp.

15.6.2.8 RDL Space Corporation

Much of the time photogrammetry deals with large-scale mapping, while the resolution of satellite data is compatible with medium- to small-scale mapping. During the recent past the sciences of GIS, GPS, and photogrammetry necessarily became comingled to satisfy the solution needs of specific projects. With regard to large-scale mapping, one of the issues arising from the utilization of satellite data has long been the matter of pixel resolution.

The RDL Space Corporation has developed a satellite system that will soon be gathering commercially obtainable information at a finer resolution. This radar imaging system, known as RADAR1, is capable of delivering SAR data to a pixel size of a single meter. As is true with all radar procedures, RADAR1 is undeterred by the time of day, weather, or clouds. By logging on to the web site http://www.rdl.com/space_corp/space_corp.html the reader may find a contact that will furnish more specific information about this satellite.

Table 15.11 Characteristics Comparison of Several Sensors

	Swath (km)	Revisit	Archive	Area
RADARSAT-1	50–500	24 days	1996	Global
ERS	100	35 days	1992	Global
LANDSAT 4,5	185	16 days	1972	Global
LANDSAT 7	185	16 days	1999	Global
SPOT	60	26 days	1986	Global

15.6.2.9 Data Comparison

Table 15.11 notes some comparable characteristics of several satellite systems.

15.7 AIRBORNE SENSORS

Some sensor systems are transported by aircraft rather than satellites.

15.7.1 Airborne Visible Infrared Image Spectrometer

One such sensor is the Airborne Visible Infrared Image Spectrometer (AVIRIS)* payload, operating as a cooperative effort by NASA (National Aeronautics and Space Agency) and JPL (Jet Propulsion Laboratory), which employs a NASA ER-2 aircraft as a platform flying at an altitude of 20 km above sea level, or an Otter aircraft at lower altitude.

Each sweep of its "whisk broom" action samples a swath 11 km wide with a pixel resolution measuring 17 m. Its various detectors sense 224 electromagnetic channels, each 0.01 μm wide, simultaneously in the 0.4- to 2.5-μm range of the electromagnetic spectrum, which covers visible, near infrared, and portions of the far infrared. Figure 15.10 represents the concept of a stack of band measurements captured in a single sweep of the scanning mirror. Simplistically, the channels in a hyperspectral capture may resemble multiple layers of information in a GIS database.

Some applications that have been used or might be considered for use with the AVIRIS system may be found in Table 15.12.

15.7.2 Thermal Infrared Multispectral Scanner

Another airborne system is the Thermal Infrared Multispectral Scanner (TIMS), developed in a cooperative effort by NASA, JPL, and Daedelus Enterprises. This optical scanner is carried by a NASA aircraft, mostly within the confines of the United States. Table 15.13 lists the spectral channels sensed by TIMS.

This variable scan-rate sensor has a resolution of 25 ft when flown an altitude of 10,000 ft. Refer to the web site http://edcimswww.cr.usgs.gov:5725/sensor_documents/tims_sensor.html for further information.

* The web site http://makalu.jpl.nasa.gov/ presents a discussion of the AVIRIS system.

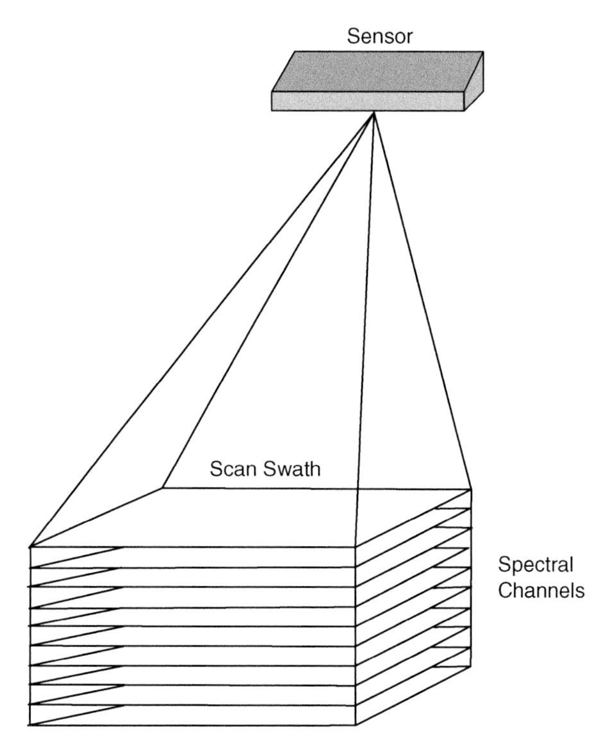

Figure 15.10 Representation of the concept of a stack of band measurements captured in a single sweep of the scanning mirror.

Table 15.12 AVIRIS Applications

Atmosphere	Snow and Ice Hydrology
Water vapor	Snow cover fraction
Clouds	Grain size
Gases	Impurities
	Melting
Ecology	
Vegetation species	Biomass Burning
Community maps	Smoke
Vegetation chemistry	Combustion products
	Fire temperature
Geology	
Mineralogy	Environmental Hazards
Soil types	Geological substrate
	Contaminants
Coastal and Inland Waters	
Chlorophyll	Commercial
Plankton	Mineral exploration
Dissolved organics	Agriculture
Bottom composition	Forest status

Table 15.13 Spectral Channel of TIMS Sensor

Channel	Spectral (μm)
1	8.2–8.6
2	8.6–9.0
3	9.0–9.4
4	9.4–10.2
5	10.2–11.2
6	11.2–12.2

Developed for use in geological studies, this system also has uses in vegetative biomass studies.

15.7.3 Digital Multispectral Videography

DMSV is the collection of multispectral data of the earth's surface from a fixed-winged aircraft with the use of a multispectral video camera system. These systems can be economical and timely data collection tools for small projects requiring analysis of land use and change similar to systems utilized in satellite-based platforms. Color Figures 1 and 2* demonstrate the collection of natural color and color infrared DMSV data over a site in Alaska. Color Figure 3* further demonstrates a thematic map generated from DMSV data. The thematic map represents supervised classification techniques used to analyze vegetation types. The use of DMSV allowed for the data to be collected at the optimum time period from a fixed-winged aircraft. This allowed the project manager to minimize typical scheduling and cost issues. The aircraft and sensor are generally under the control of the mapping firm. The necessary flights are therefore scheduled at the optimum time periods on relatively short notice. Flight cost may be minimized due to the economics of using a small aircraft and support staff along with the fact that the aircraft mobilization time to the project site may be kept to a minimum.

15.8 SOURCES OF SATELLITE IMAGERY

Scanned data and image reproductions are available for purchase.

15.8.1 SPOT

Information concerning SPOT satellite products can be obtained from:

SPOT Image Corporation
1897 Preston White Drive
Reston, VA 20191-4368
Tel: (703) 715-3100 or (800) ASK-SPOT
Fax: (703) 648-1813

* Color figures follow page 42.

The web site http://www.spot.com/home/distri/na/List_2000.htm lists the North American distributors of this product.

15.8.2 LANDSAT

Information concerning LANDSAT satellite products can be obtained from:

Space Imaging EOSAT
12076 Grant Street
Thornton, CA 80241
Tel: (800) 232-9037

15.8.3 National Oceanic and Atmospheric Agency

Information concerning NOAA satellite products can be obtained from:

Satellite Data Service Division
National Climatic Data Center
World Weather Building, Room 1
Washington, D.C. 20233

The Emporia State University in Kansas provides a web page that may be of interest to the potential user of geospatial data. Log on to http://www.emporia.edu/earth-sci/geosourc.htm for a listing of 38 on-line sources for global GIS and geospatial data, primarily at governmental agencies and universities.

15.8.4 Space Imaging EOSAT

IRS, ERS, IKONOS-1, and LANDSAT imagery may all be ordered from:

Space Imaging EOSAT
12076 Grant Street
Thornton, CO 80241
Tel: (800) 232-9037

CHAPTER 16

Image Analysis

16.1 ANALYSIS PROCEDURES

Studies for various disciplines require different technical approaches, but there is a generalized pattern for geology, soils, range, wetlands, archeology, hydrology, agronomy, forestry, habitat, land use, or other similar study purposes.

Information system databases can be multidisciplinary, and several studies can run concurrently. The combined information may be correlated so as to form the database necessary for a multifaceted GIS study. Prior to beginning a project the objectives of the study must be defined. In the past, the analysis and mapping work was accomplished by manual methods, mostly utilizing monoscopic aerial photographs. Since this procedure does not remove inherent object displacement, the accuracy of the mapping may have been suspect on intensive studies in areas exhibiting terrain relief. Softcopy mapping has revolutionized these procedures, since this mapping methodology depends upon stereoscopic procedures where the integration of vertical data creates an orthogonal base map.

The progression of an image analysis investigation may include the techniques discussed in this chapter.

16.1.1 Data Acquisition

Project managers must acquire appropriate information packages pertinent to the project site, such as:

1. Aerial photographs and/or digital data from multispectral scans, thermal scans, radar imagery, or a combination of several of these sources
2. DEM or DTM or LIDAR elevation point data to generate a vertical base
3. Cadastral data relevant to rectifying the digital information to its true geographic scene
4. Tabular and textual material germane to analyzing the created thematic polygons

16.1.2 Rectify Data

Upon merging the digital image information with the vertical database and the geographical datum in a softcopy machine, electronic mapping can proceed.

16.1.3 Thematic Map Generation

Specific classification themes relative to the prime discipline (soil, crop, forest, habitat, wetlands, hydrology, geology, urban, land use, and others) are thematically classified. Currently, the trend is to achieve this through electronic assistance involving computers and complex software. The use of interpretation keys may be of use to carry out this task.

Computer hardware, peripheral facilities, and software capable of storing, manipulating, and outputting the vast amounts of generated data are costly. Potentially, these data analysis systems can be of great value to the image analyst in various scientific specialties relative to capturing GIS data from photographic images exposed with aerial or metric cameras and raster data collected by remote sensors.

With judicious guidance from the technician, these systems are capable of classifying themes, measuring acreage, creating contours and shaded relief, producing aspects and perspectives, and gleaning other information from the images automatically. Utilized properly, softcopy mappers allow the scientist to reduce time spent on mundane labor-intensive processes so that quality time can be devoted to matters requiring professional acumen.

The electronic classification procedures create a thematic map automatically. Satellite imagery in digital form is especially useful in GIS studies that dictate small-scale cartography. Multiple scenes can be mosaicked to form a single composite.

It is possible to automatically create theme maps (water, urban, woodlands, crops, wetlands, soils, fallow) from photographs or images generated by digital data in at least two fashions: supervised and unsupervised classification.

16.1.3.1 Supervised Classification

In supervised classification the technician selects training sites, which are sample areas representative of specific thematic signatures on the image. Sample signatures can be based on ground truth, interpretation keys, or some other knowledge of the particular theme. The computer will then select other areas of similar brightness value and paint these on the viewing screen. This is a valuable system characteristic if the user has access to valid knowledge of the various thematic classes. Interpretation keys may be helpful in selecting training sites. Refer to Color Figure 3* for an example of a thematic map created by supervised classification procedures.

16.1.3.2 Unsupervised Classification

In unsupervised classification the technician requests the computer to make a thematic segregation. The computer then breaks the range of reflectances into the

* Color figures follow page 42.

number of separate classes suggested by the technician and paints areas of this group of brightness value ranges on the screen. Once these computer themes are available, the operator must devise a method such as field investigation or low-altitude aerial photography to identify the computer-selected classes.

16.1.4 Areal Mensuration

Once the thematic separation is made, by whichever process, the system can measure areas and apply these acreages to other information appropriate to the project. Perimeters, percentages by class, and other pertinent mensurational data are also electronically extracted from the classification map.

16.1.5 Ground Truth Sampling

Field visitation is necessary to visually verify the integrity of the image classifications. Collecting ground truth information from sample areas scattered throughout each class may entail considerable temporal and monetary expenditures. In fact, this field procedure could require more effort than that required to complete office routines.

16.1.6 Data Correlation

Areal measurements within a theme are correlated with the relevant ground truth data and tabular information gathered at the ground sampling stations in order to assess the characteristics of the total project area.

16.1.7 Reporting

Project reports are prepared, often with a great deal of input from the computer, to include graphs, charts, tables, maps, etc.

16.2 IMAGE INTERPRETATION KEYS

Image interpretation keys are graphic and/or textual aids that may help identify thematic classes from image feature characteristics. Some keys may be available from varied sources. Sometimes it is necessary, at the beginning of a project, to construct a key to be employed in that specific or in similar studies. The technician, when selecting class target samples in supervised classification, should keep in mind the concept of interpretation keys, because that is the basis for this procedure. Not all objects can be identified by use of an interpretation key only. Photo interpreters must possess an intuitive deduction trait.

16.2.1 Composition of Keys

Interpretation keys are usually comprised of one of two methods of presenting the discriminating information: stereograms and descriptive text. Some keys may combine both.

Interpretation keys may be composed of a collection of stereograms, which are pertinent portions of stereopairs which when viewed in three dimensions form graphic illustrative samples of various objects.

Interpretation keys may be composed of word descriptions defining the characteristics of various features.

16.2.2 Types of Keys

Basically, interpretation keys can be one of two generic types: selective and elimination.

Selective keys are used by the analyst to search through the stereograms and/or word descriptions until a match is made with the photo image object.

Elimination keys are composed of word descriptions ranging through various levels of broad to specific characteristic discrimination. The analyst progresses down through this hierarchy, making choices at branching description paths. Finally, by the process of eliminating all differing features, the object is identified.

16.2.3 Interpretation Characteristics

A number of image characteristics are scrutinized in order to identify ground objects.

16.2.3.1 Shape

The form or configuration of an object can eliminate or identify objects. Racetracks are seen as elongated ovals on the image. Not too many objects exhibit this particular shape characteristic, other than athletic tracks and raceways.

16.2.3.2 Size

The relative size of an object can help to discriminate features. A soccer field and a tennis court can exhibit somewhat similar rectangular signatures on a photo. The standard dimensions of these features are significantly different. By measuring the size of these features on the image, the discrimination is apparent, as would be the difference in size between a residence and a doghouse.

16.2.3.3 Pattern

Spatial arrangements can definitely aid in identifying objects. Both orchards and woodlots are composed of trees. The trees in a natural woodlot are situated randomly and, since the individual trees are not all the same age, attain various heights. Orchards and plantations are planted with grids of evenly spaced trees that are all, since they are planted at the same time, roughly the same height.

16.2.3.4 Shadows

The characteristic forms of some features are difficult to identify, except by the shadows they cast. A utility pole is sometimes difficult to see on the image, but it can be more easily located by its shadow.

16.2.3.5 Tone

The relative brightness of an object's spectral signature can help identify that feature. The highly reflective surface of a sand trap on a golf course may appear lighter on the image. Conversely, a putting green reflects limited light and will appear darker on the image.

16.2.3.6 Texture

The frequency of tonal change of an object's surface may give an indication of its identity. The varying diameters and heights of individual plants in a soybean field create a "cobbled" effect on an image. On the other hand, a pasture is composed of relatively short grass which exhibits an even texture on the image.

16.2.3.7 Site

One can expect to find objects in certain environmental situations. Grain fields are not cultivated in marshes, yet tracts of wild rice may be found there.

CHAPTER 17

Project Planning and Cost Estimating

17.1 INTRODUCTION

Previous chapters have outlined and detailed technical aspects of photogrammetry. The basic tasks and equipment required to create various mapping products have been clearly noted. These tasks generally include aerial imagery collection, ground survey control collection, feature collection and attribution, and formatting the final spatial data. Further, the details regarding necessary equipment and product accuracy have been highlighted. Successful project planning and management must include a thorough understanding of these details. This chapter will provide guidance for the development of a set of specifications and a methodology to develop budgetary estimates of effort and cost to produce typical photogrammetric mapping data sets.

17.2 SPECIFICATIONS

Detailed project planning is the essence of an accurate effort and cost estimate. A quality specification need not be overly prescriptive, but it must include certain details: a description of what the spatial data collection is to be used for, a basic understanding of the required tasks and final products, and what will be required to define and collect them. Figure 17.1 shows a flow chart providing an overview of these items. See Appendix A for a sample specification text along with an estimate for time and cost.

17.2.1 Project Description and Boundary

The project manager should briefly describe the intended uses for the spatial data to be collected. Engineering and design mapping may require some different collection methods than a product that is to be used for GIS environmental analysis.

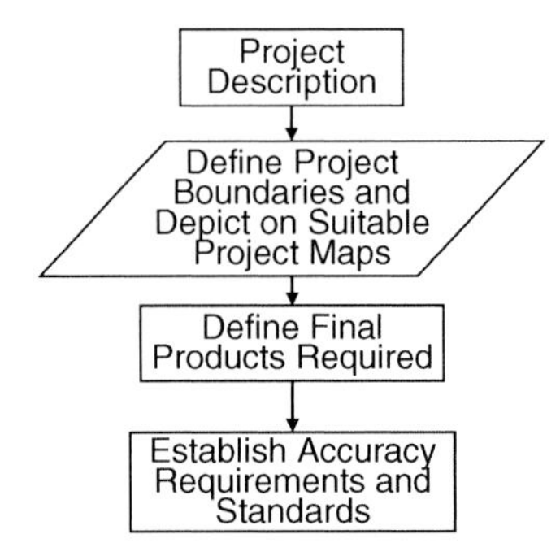

Figure 17.1 Flow chart providing an overview of quality specification for the required tasks and final products and what will be required to define a project design.

Table 17.1 Suggested Spatial Data Purposes vs. Map Scales

Intended Use of Spatial Data	Suggested Horizontal Map Scale	Suggested Contour Interval
Regional and general planning	1 in. = 1,000 ft to 1 in. = 2,000 ft (1:12,000 to 1:24,000)	10–20 ft
Environmental and real estate studies	1 in. = 400 ft to 1 in. = 800 ft (1:4,800 to 1:9,600)	4–8 ft (1.5–2.5 m)
Citywide and facility management studies	1 in. = 100 ft to 1 in. = 200 ft (1:1,200 to 1:2,400)	2–4 ft (0.75–1.5 m)
Detailed engineering design and layout	1 in. = 30 ft to 1 in. = 50 ft (1:300 to 1:600)	1–0.5 ft (0.5–0.2 m)

EXAMPLES

This photogrammetric mapping project is intended to be used for the facilities management and GIS analysis of the Oakville Industrial Complex. The project will include an area-wide planimetric and topographic mapping database along with digital orthophotography.

Large-scale mapping (1 in. = 100 ft to 1 in. = 30 ft) will require different photography scales and survey efforts in addition to a different feature collection criteria than small-scale mapping (1 in. = 1200 ft to 1 in. = 2000 ft). Table 17.1 will assist in making a decision regarding an appropriate map scale.

The project boundary (limits) should be clearly defined. It is best to define the boundary as a line on a suitable published map with a measurable scale.

The project boundary is annotated by the polygon drawn on the copy of the USGS 7.5 ft Warrenton County, MO quadrangle. This project boundary map is considered part of the specifications for this project.

A USGS 7.5-ft quadrangle sheet is a common base map for a project boundary map. These map sheets are published at a scale of 1:24,000 (1 in. = 2000 ft) and are easily obtained at minimum cost from the USGS or other public sources. Projects involving large areas (i.e., projects that may cover portions of several USGS 7.5-ft quadrangle maps) may be depicted on other USGS map products such as 1:250,000 or 1:1,000,000 scale sheets.

The boundary map should provide a polygon indicating the maximum mapping limits. Stipulation of a buffer zone beyond the mapping limits should be avoided. Photogrammetric mapping firms generally provide a minimal buffer beyond the mapping polygon to account for any discrepancies due to inaccuracies in the base map and/or in defining the boundary polygon. However, the project manager should not count upon this buffer. A vague boundary leads to difficulty in estimating project requirements and costs. Some projects may call for a photography boundary that is larger than the mapping boundary. Additional boundaries such as this should also be clearly marked on the project map. Figure 17.2 shows a typical boundary map.

17.2.2 Define Products

Clear definition of the products required for a project is mandatory and should be clearly understood in the initial stages of project planning and cost development. The costs and overall time required are dependent upon the delivery item requirements. Products may be divided into two groups: intermediate and final.

1. Intermediate products would include items such as:
   - Processed aerial film
   - Paper prints
   - Diapositives and/or scanned images
   - Ground survey control data
   - Project planning reports
   - Aerotriangulation reports
   - Check map sheets
2. Final products would include such items as:
   - Paper prints of aerial film
   - Hardcopy map sheets
   - Digital planimetric and/or topographic map data
   - Digital orthophotograph

EXAMPLE

The final deliverables for this project will include two sets of black and white 9 × 9 in. prints and one set of digital planimetric and topographic maps at 1 in. = 100 ft with 2-ft contours on CDROM.

17.2.3 Standards and Accuracy Requirements

Stating a required horizontal map scale and contour interval for a map is not sufficient alone to establish effort and cost. To truly design and estimate the effort required in collecting and presenting a set of spatial data products, horizontal and

Boundary

Figure 17.2 Sample project boundary map.

vertical accuracies must be stated. Several standards have been discussed in previous chapters, namely, FGDC, NMAS, and ASPRS. All three are widely accepted today. However, NMAS and ASPRS have been around for a longer period of time, and for some casual spatial data users they are more familiar. Regardless, it is imperative in a successful specification to not only state the map scales required, but also to state the accuracy standard to which they will be held.

EXAMPLE

The planimetric and topographic mapping data produced for this project will meet ASPRS Class I accuracy standards for 1 in. = 100 ft with 2-ft contours.

17.3 PROCESSES

As previously stated, the project manager must understand the basic processes required to produce the final products requested. This understanding is necessary to ensure that the specifications and cost estimates include the necessary intermediate products required in the production of the final products and to evaluate and compare proposals. A project manager must be careful not to be too prescriptive in describing these processes so as to exclude new or the most efficient technology. However, some intermediate products may have other uses. For example, control survey points may be used for other projects such as construction of structures. It may then be necessary to request that the ground control be established as a permanent benchmark with several reference points and a concrete post with a descriptive cap on top. These types of requirements need to be spelled out in a set of specifications. However, the type of equipment to be used (brand and model) is not necessary.

EXAMPLE

Ground survey control points shall be collected for photogrammetric mapping. The mapping contractor will develop a ground survey plan that will produce the final mapping at the specified accuracy.

The processes for a typical photogrammetric mapping project generally include:

- Aerial imagery collection
- Ground survey control collection
- Densification of ground control through the use of aerotriangulation procedures
- Elevation data collection
- Planimetric feature data collection
- Final formatting of data sets

These functions should be noted in a set of specifications in order to develop an independent budget estimate of the time and funding required and to allow the project manager to truly compare proposals from several prospective contractors.

17.3.1 Aerial Photography

Aerial photography should indicate the following:

- Type of camera system (analog or digital)
- Type of film or sensor (black and white, natural color, color infrared, etc.)

- Negative scale of the imagery (i.e., 1 in. = 300 ft or 1:3600)
- Focal length and format of the camera (i.e., 6-in. focal length, 9 × 9 in. format aerial camera)
- USGS camera calibration within the time period of the last calibration (i.e., three years)
- Approximate altitude above mean terrain (AMT)
- Number of sets of paper prints and or scans

17.3.2 Ground Survey Control

The ground survey control process should note the following:

- Procedures that will produce survey locations with accuracy suitable to produce the final photogrammetric mapping reliability shall be listed.
- A ground survey plan shall be developed that will indicate control monuments used in the project, the approximate location of the proposed ground control points, data to be used for the project (both vertical and horizontal datums), and the proposed methods that will be used to establish their respective locations.
- Ground control point location information will then be tabulated, recorded, and marked on one set of paper prints (control prints) for future mapping.
- A final ground control report shall be prepared that compiles the results of the survey along with a narrative describing the procedures used, any problems encountered, and how they were resolved.

17.3.3 Densification of Ground Control

This process generally is accomplished through the use of aerotriangulation (AT) processes. The specifications should note whether analog or softcopy AT processes will be used. Analog AT will require the production of diapositives, and softcopy AT will require the production of scanned images suitable for softcopy AT. An AT report should be generated that indicates the methods used, expected results, final results, any problems encountered, and how they were resolved.

17.3.4 Elevation Model Collection

Elevation model collection guidance should define the purpose of the elevation model. Elevation models generated for contour production require more detail than those used only for orthophoto rectification. Elevation models for contour generation are generally produced by collecting mass points and breaklines to adequately define the surface. A surface model or TIN may also be required for software to use in the generation of contours. It is also very important to state the expected contour interval and accuracy standard.

17.3.5 Planimetric Data Collection

Planimetric feature data collection guidance may be divided into two categories.

Features are normally collected from aerial photo negative scale for a specific map scale. As a general rule, the larger the photo negative and map scale is, the more detail that can be seen, identified, and plotted. Generally, the specifications state that the photogrammetric technician shall plot all features visible and identifiable in the imagery.

However, some projects may require limited and/or very specific planimetric feature data collection (i.e., only water boundaries and major roads). Cases such as these may require that the specifications list the features to be collected and/or note the features not to be collected.

17.4 ESTIMATING PRODUCTION EFFORT AND COST

Project managers should prepare an independent estimate of the time and costs associated with a photogrammetic mapping project. This independent estimate will help assure that the products requested will be purchased at a fair and reasonable cost and will also allow the project manager the knowledge required to compare proposals and determine differences between them. These independent estimates must account for all significant phases of the project. A process is described in the following sections that will allow a project manager familiar with the basic processes as described in previous chapters in this text to develop a budget estimate of time and associated cost for a typical photogrammetric mapping project. New technology and unique project requirements may make this process not applicable. However, this section will present a generalized protocol that may help to guide cost estimates of most projects that require aerial imagery along with planimetric and topographic mapping that can fit the processes, and if followed closely a reasonable budgetary estimate of time and cost can be developed.

17.4.1 Estimating Factors

Three major considerations must influence the project's cost estimate: labor, equipment, and overhead and profit.

17.4.1.1 Labor

The involvement of qualified scientists and technicians is one of the most significant factors in the project cost. An accurate budget should estimate the time and unit cost for each required function. The amount of work required by various staff members is considered direct labor. For the purposes of this chapter, the staffing effort will be expressed in hours along with a unit cost per hour. Labor rates for most of the disciplines listed in this chapter can be obtained from local labor rate schedules and/or by consulting with other project managers in the area who have had similar work accomplished for them.

17.4.1.2 Equipment

Hardware and ancillary software associated with the various staff operations are also significant cost items to consider. As stated in earlier chapters, equipment such as aircraft, cameras, stereoplotters, and scanners can be very costly. The project manager should consult others who have had similar work completed recently for hourly costs for these items.

17.4.1.3 Overhead and Profit

Although these significant items may tend to be subjective, they must be considered as objectively as possible. Overhead and profit for photogrammetric mapping firms may legitimately be higher than many other engineering firms in the vicinity. As stated above, maintenance and updating of equipment, as well as the necessary staff to operate and maintain it, are expensive. Aerial photography and ground surveys are often affected by weather and access at the project site, so the risk may be greater than for some other design projects. Therefore, profit percentages may be more than those normally associated with other design projects.

17.4.2 Costing Aerial Photography

Many factors, including manpower and equipment rental, must be considered in estimating production hours in the aerial photo mission.

Production Hours for Aerial Photography

Direct Labor
- Project mission:
  - Flight preparation = 1.5 h
  - Takeoff/landing = 0.5 h
  - Cross-country flight = miles to site × 2 ways/mph
    - = ______ × 2/______
    - = ______ h
  - Photo flight =
  - End turns = lines × 0.08 h = ______ h
- Photo lab:
  - Develop film = ______ photos × 0.04 = ______ h
  - Check film = ______ photos × 0.04 = ______ h
  - Title film = ______ photos/40 = ______ h
  - Contact prints = ______ photos/45 = ______ h

Equipment Rental
- Aircraft = project mission hours = ______ h
- Airborne GPS = project mission hours = ______ h (if not included in aircraft rental)
- Film processor = develop film hours = ______ h
- Film titler = title film hours = ______ h
- Contact printer = contact prints hours = ______ h

17.4.3 Costing Photo Control Surveying

The specified accuracy requirements will dictate methods and procedures (conventional and/or GPS) to be employed as well as indicate the number and pattern of control points to be established. A pre-project field survey plan should be conceived and thoroughly discussed with a licensed surveyor.

Photo control is often estimated either as an average cost per control point or with a detailed staffing and equipment estimate for the total project. In these cases it is usually fair to estimate these efforts and cost on a per control point method. Detailed estimates should be reserved for unique projects where the ground surveys will compose a large part of the total project cost. The project manager should consult with others who have obtained these products to get a current unit cost that is fair and reasonable.

17.4.4 Costing Aerotriangulation

Many factors, including manpower and equipment rental, must be considered in estimating production hours in the AT procedure.

Production Hours for Aerotriangulation

Direct Labor
  Photo scan = ________ photos × 0.3 h = ______ h
  Aerotriangulation (workstation):
    Model orientation = ______ models × 0.2 h
      = ______ h
    Coordinate readings = ______ photos × 0.3 h
      = ______ h
  Computations = ______ models × 0.4 h = ______ h

Equipment Rental
  Scanner = scanning hours = ______ h
  Workstation = aerotriangulation hours = ______ h
  Computer = computations hours = ______ h

17.4.5 Costing Photogrammetric Compilation

For site-specific information, the following items are to be calculated, estimated, or measured to assist in the computing costs associated with digital mapping:

1. Number of stereomodels to orient
2. Number of acres to map
3. Complexity of terrain character
4. Complexity of planimetric culture
5. Format translations of digital data

Production Hours for Stereomapping

Model Setup

Model setup includes planning the collection procedures and georeferencing models in the data collection system. Analytical stereoplotters or softcopy workstations may accomplish data collection. Analytical stereoplotters will require diapositives, and softcopy workstations will require high-resolution scans. For additional explanation and detail, review portions of Chapters 5–10.

Model orientation = _______ models × 0.1 hours = ______ h

Photo scan = ________ photos × 0.2 hours = ______ h
(if not done previously)

Digital Data Capture

Planimetry (cultural features)

The project planning map used to outline the mapping area should be overlain with a proposed flight line layout. The flight line layout should note the approximate location of each photo stereopair. The planimetric feature detail in each of the models should be assessed based on the amount of planimetric detail to be captured (full or partial stereomodel and the final map scale) and the density of planimetry to be captured in each stereomodel. For example, highly urban area stereomodels require more time to compile than rural area stereomodels. Table 17.2 will aid in estimating the production time required to produce the planimetric features.

Topography

The project planning map used to outline the mapping area should be overlain with a proposed flight line layout, which should note the approximate location of each photo stereopair. The topographic feature detail in each of the models should be assessed based on the amount of topographic detail to be captured (full or partial stereomodel and the final map scale). Topographic detail must consider the character of the land to be depicted. For example, a 1-ft contour development in a relatively flat terrain requires much less time than collection of 1-ft contours in very mountainous terrain. Table 17.3 will aid in estimating the production time required to produce the topographic features.

17.4.6 Costing Orthophoto Images

Current technology allows for total softcopy generation of orthophotos (see previous chapters for more detailed information). If a contractor has collected the digital terrain mode with an analytical stereoplotter and created diapositives, then a clean set of diapositives will need to be made and scanned for orthophoto generation. However, if the contractor uses softcopy stereocompilation for the elevation model collection, then the same scanned images may be used to generate the orthophotos. The user must assume one method or the other in developing a cost estimate. The difference in cost should be negligible.

Table 17.4 is a chart for estimating the production time required to produce topographic features.

17.4.7 Summary of Production Efforts

Table 17.5 provides a summary of the production hours itemized above. Note that in addition to the total labor hours, an appropriate overhead should be established and applied to the total cost of labor. Also, an appropriate profit should be established and applied to the total of labor and direct costs. Ground survey requirements

Table 17.2 Chart for Estimating Production Time Required to Produce Planimetric Features

Planimetry				Approximate Planimetric Time (Hours/Model)			
				Final Map Horizontal Scale			
Density Type	Models Per Type	Hours Per Type	Total Planimetry Hours	1 in. = 40 ft to 1 in. = 60 ft	1 in. = 100 ft to 1 in. = 150 ft	1 in. = 200 ft to 1 in. = 300 ft	1 in. = 400 ft to 1 in. = 1600 ft
Light planimetry							
1				3.0	2.5	2.5	2.5
2				4.0	3.5	3.5	3.5
Medium planimetry							
3				5.0	4.0	4.0	4.0
4				7.0	6.0	6.0	5.0
Heavy planimetry							
5				10.0	8.0	7.0	6.0
Total planimetry hours							
Edit time: generally 30% of total planimetric compilation hours							

Table 17.3 Chart for Estimating Production Time Required to Produce Topographic Features

Topography (Topo) Collection of Mass Points and Breaklines for Production of Contours				Approximate Topography Time (Hours/Model)			
				Final Map Contour Interval Scale			
Terrain Character (Slope)	Models/Type	Hours/Type	Total Topo Hours	1 ft	2 ft	4 ft	5–8 ft
Flat				2.0	2.5	2.5	2.0
Rolling				4.0	4.0	4.0	3.0
Hilly				6.0	6.0	5.0	4.0
Steep				8.0	8.0	6.0	5.0
Disturbed				10.0	10.0	8.0	7.0
Total topo hours							
Edit time: generally 30% of total topo collection time							

Table 17.4 Chart for Estimating Production Time Required to Produce Orthophotographs

Orthophoto Production Costs (Direct Labor)		
Elevation Model (DEM) Development (Ortho Only) Developed by the Stereocompiler		
# Stereo Models	**Hours/Model**	**Total DEM Time (Stereo Models × Hours/Model)**
	2 h/model	

Tasks Below Accomplished by Softcopy Technician				
	Natural Color and Color IR		**Black and White**	
	Hours/Image	**Total Hours**	**Hours/Image**	**Total Hours**
Image scanning	0.3			
DEM — scan data merge	0.5			
Radiomenteric correction	2.5			
Tiling/sheeting	0.25			
Total hours				

Table 17.5 Summary of Total Estimated Production Time and Costs for the Project

Photogrammetric Mapping Project Production			
	Hours	**Unit Cost**	**Total Cost**
Production labor			
Aerial photography			
Aerotriangulation			
Model setup			
Planimetry			
Topography			
Orthophotography			
Total			
	Units	**Unit Cost**	**Total Cost**
Direct costs			
Film		Foot	
Prints		Each	
Diapositives		Each	
Hardcopy prints		Each	
CDs, disks, or tapes		Each	
Aircraft w/camera		Hours	
Stereoplotter		Hours	
Softcopy workstation		Hours	
Edit workstation		Hours	
Scanner		Hours	
Total direct cost			

established should also be added to the total costs below. Table 17.5 provides a summary of the total estimated production time and costs for the project.

By applying wage rates to labor hours and equipment rental to equipment hours the estimator can arrive at project budgetary costs. To support the process of estimating production hours of various photomapping phases, the estimator of a project is urged to seek current labor and equipment rates more applicable to a project or geographical area, perhaps using regional wage rates and actual negotiated contractor rates from recent or current contracts.

APPENDIX A

Example of a Typical Photogrammetric Mapping Project Cost Estimation

A.1 SPECIFICATION

Project Description

The contractor is requested to supply all necessary equipment, staff, and expertise required to generate full planimetric and topographic mapping of a site in Jersey County, IL, near the town of Elsah, IL. The data produced for this project will be used for general planning and preliminary engineering for flood control and infrastructure development (cut and fill). The final mapping products will include 1 in. = 100 ft planimetric mapping with 2-ft contours. The mapping will be referenced to the Illinois (East) SPCS, NAD 83, and NAVD 88. The mapping will be in feet. The site location and project boundary is shown on a portion of the Elsah, IL-MO USGS 7.5-ft quadrangle sheet (1991). A copy of the site location map with the boundary is attached to this specification. The contractor will be required to collect new black and white aerial photography, establish necessary ground control for aerotriangulation and subsequent photogrammetric mapping, perform aerotriangulation procedures necessary to densify the ground control and to rectify the imagery to the earth, collect all planimetric detail that is visible and collectable at the required map scale, and collect topography of the site in the form of mass points and breaklines, and contours. This project shall follow standard industry procedures, and the final products will meet or exceed ASPRS Class I standards for 1 in. = 100 ft mapping with 2-ft contour intervals.

Client-Furnished Information

The client has furnished as part of this specification a project location map. The project location map is a paper copy of the USGS 7.5-ft Elsah, IL-MO quadrangle sheet. The project boundary is clearly annotated on the project location map. Any additional information, right of entry, equipment, or data is the responsibility of the contractor.

Description of Tasks

The contractor will:

1. Develop a project plan that will include flight lines, ground control locations for the project, and a brief text describing the project location (including a map with the project boundary, flight lines and photo frame locations, and ground control locations). This plan will note the scale of the photography, the type of film, the forward lap and sidelap of the photography, the horizontal and vertical datums to be used for the ground control and photogrammetric mapping products, and an anticipated time line for completion of the project. The project plan will also include a brief description of quality control procedures that will be used by the contractor to validate the accuracy of the final mapping products.
2. Establish all necessary horizontal and vertical ground control for the project. Ground control may be a combination of ground panels and photo-identifiable features. Photo-identifiable features will require location data to be established after photography is completed. All ground control points shall be referenced and tied to at least two other features near each point site. A neat sketch of each site describing the point, its location, and the location of the tie points shall be prepared. A ground control report shall be prepared describing the ground control plan, control points used, expected accuracies, and final accuracies. This report, to be signed and stamped by a registered land surveyor of the state of Illinois, will also provide a map indicating the location of the actual points (a copy of the 7.5-ft USGS quadrangle) and control points used. Any problems encountered and how they were resolved will be discussed in the report.
3. Fly and photograph the site with black and white film during leaf-off conditions during the early spring of the year. The photography will be captured with minimal cloud cover (less than 5% in any frame), no snow on the ground, and no flood waters that would obscure ground information collection. Aerial photography shall be collected during a period of the day when the sun angle is 30% or higher and captured at an approximate photo scale of 1 in. = 500 ft with a forward lap of 60% and sidelap of 30%. The camera used shall be a typical 9 × 9 in. format metric aerial photography camera with a 6-in. focal length lens. The camera shall have a current (within the last three years) USGS certification. A copy of the USGS certification shall be furnished as part of the final product for this project. The film will be processed and labeled, and two sets of paper black and white prints (9 × 9 in.) will be produced of each exposure. Film labeling shall be across the top of each exposure with the date of photography, project name (Elsah, IL), photo scale (1 in. = 500 ft), flight line, and frame numbers.
4. Mark the ground control locations on the back of one set of prints to be used for aerotriangulation and mapping. The location and type of control point (horizontal and/or vertical) shall be marked on the front of required control prints.
5. Generate diapositives or scanned images to be utilized in the aerotriangulation process and subsequent map feature compilation.

6. Utilize the ground control and diapositives with appropriate software and hardware to generate a suitable aerotriangulation process that will allow map compilation that will meet or exceed ASPRS Class I standards for 1 in. = 100 ft mapping with 2-ft contours.
7. Generate an aerotriangulation report that will include the procedures, software, and hardware used in the aerotriangulation effort. This report will indicate the expected accuracy of the final aerotriangulation process, as well as the results of the process, and will discuss any problems encountered and how they were resolved, including ground points withheld from the solution, why they were withheld, and how this affected the final solution. The report will be signed by the author and the project manager.
8. Employ either softcopy or analytical stereoplotter methods to collect the planimetric features within the project boundaries. Feature collection will follow and be in compliance with the FGDC standards. All planimetric features that can be seen and plotted shall be collected. Feature collection will include, but is not limited to, all roads, trails, buildings, permanent structures, bridges, utility poles, edges of water bodies, dams, walls, parking lots, tanks, silos, sporting facilities, cemeteries, levees, aboveground pipelines, and airport facilities.
9. Collect topographic features (in ASCII format) throughout the project area, which includes mass points and breaklines and contour files that will describe the character of the earth's surface within the project boundary. In addition, the topographic detail in the contour files will note areas of major high and low points as spot elevations. Sufficient topographic detail in flat areas will be collected and displayed to depict the general lay of the land.
10. Provide the final data sets on CDROM disks. Two copies of planimetric data and contour files will be submitted in AutoCad Version 14, and the mass points and breakline files will be submitted in the ASCII format that is fully compatible with AutoCad Version 14.
11. Produce metadata for the entire project to include the aerial photography, ground control, and all feature collection that is fully compliant with the FGDC "Content Standard for Digital Geospatial Metadata," FGDC-STD-001-1998.

Deliverables

The final deliverables will include:

- A project plan
- All exposed film
- Two sets of prints (one clean set and one control set)
- One copy of the USGS camera calibration report for the cameras used for the project
- All ground control information and ground control reports
- Aerotriangulation report
- Two sets of final data on CD; final data sets include planimetric features in AutoCad version 14, mass points and breaklines, and contour files
- One digital set of the FGDC compliant metadata

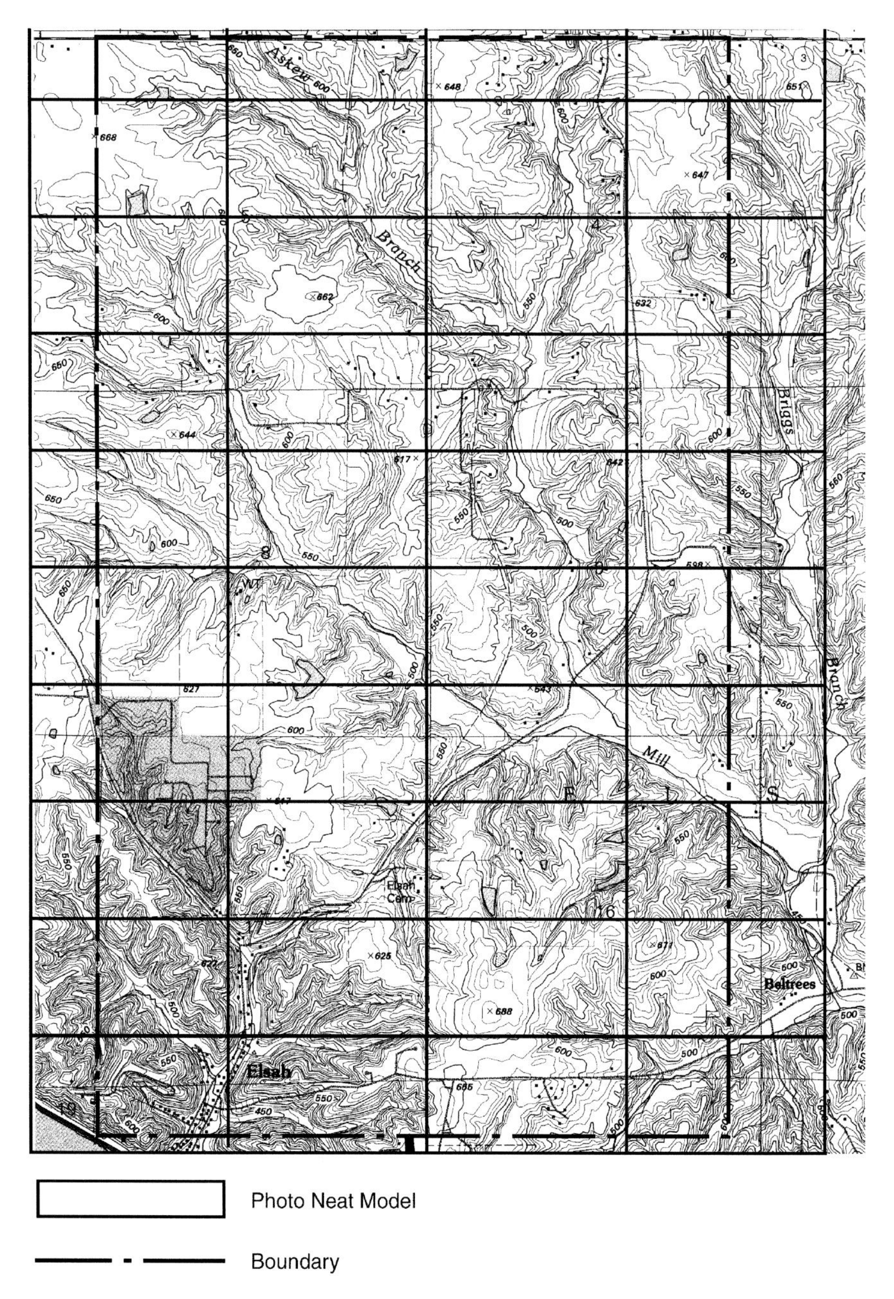

Figure A.1 Photo neat model layout.

A.2 COST ESTIMATION PROCESS

The project described above is briefly mentioned and described in Chapters 9 ("Photo Scale Selection") and 11 ("Aerotriangulation"). The project boundary map (including approximate ground control and flight line locations) noted in the specifications is shown in Figure 11.4. Figure A.1 is another map depicting the approximate layout of photo neat models over the project area for the prescribed photography. As noted in Chapter 9, photo coverage for this project will require approximately 44 photos (4 flight lines with 11 photos per flight line). This total number will include one additional photo per flight line to insure complete stereo-coverage of the project area. Figure A.1 shows a layout of stereo neat models that will be required for compilation only. This layout will be the basis for the estimates described below. The estimates described below follow the procedures in Chapter 17, "Planning and Cost Estimating." Refer to this chapter for additional detail.

Production Hours for Aerial Photography

Direct Labor
- Project mission
  - Flight preparation = 1.5 h
  - Takeoff/landing = 0.5 h
  - Cross-country flight = miles to site × 2 ways/mph
= 50 × 2/200
= 0.5 h
  - Photo flight = 0.5 h to collect the photos over the flight lines
  - End turns = 4 lines × 0.08 h = 0.5 h
- Photo lab
  - Develop film = 44 photos × 0.04 = 1.8 h
  - Check film = 44 photos × 0.04 = 1.8 h
  - Title film = 44 photos/45 = 1 h
  - Contact prints = 88 photos/45 = 2 h

Equipment Rental
- Aircraft = project mission hours = 2 h
- Film processor = develop film hours = 1.8 h
- Film titler = title film hours = 1 h
- Contact printer = contact prints hours = 2 h

Photo control is often estimated either as an average cost per control point or with a detailed staffing and equipment estimate for the total project. As a general rule, photo survey control is not a large cost contributor. In these cases it is usually fair to estimate these efforts and cost on a per control point method. Detailed estimates should be reserved for unique projects where the ground surveys will compose a large part of the total project cost. The project manager should consult with others who have obtained these products to get a current unit cost that is fair and reasonable. For this example we are assuming that the ground control will be as described in Chapter 11 and shown in Figure 11.4. Approximately 20 horizontal/vertical points will be required. We will also assume that the collection and processing of all required ground control information as described in the specifications above will require approximately three days for a three-person survey crew plus two days in the office checking data and producing necessary reports. The cost for this effort will be $12,000. Note that these numbers are intended for this example only.

Production Hours for Aerotriangulation

Direct Labor
  Photo scan = 44 photos × 0.3 h = 13.2 h
  Aerotriangulation (workstation):
    Model orientation = 44 models × 0.2 h
      = 8.8 h
    Coordinate readings = 44 photos × 0.3 h
      = 13.2 h
  Computations = 44 models × 0.4 h = 17.6 h

Equipment Rental
  Scanner = scanning hours = 13.2 h
  Workstation = aerotriangulation hours = 22 h
  Computer = computations hours = 17.6 h

It is assumed that this project will be compiled with softcopy methods. Therefore, diapositives will not be required for aerotriangulation or compilation. The film will be scanned directly, and the scanned images will be incorporated into the system.

The following items are to be calculated, estimated, or measured to assist in the computing costs associated with digital mapping:

1. Number of stereomodels to orient
2. Number of acres to map
3. Complexity of terrain character
4. Complexity of planimetric culture
5. Format translations of digital data

Production Hours for Stereomapping

Model Setup

Model setup includes planning the collection procedures and georeferencing models in the data collection system. Data collection may be accomplished by analytical stereoplotters or softcopy workstations. Analytical stereoplotters will require diapositives, and softcopy workstations will require high-resolution scans. For additional explanation and detail, review portions of Chapters 5 through 12.

  Model orientation = 40 models × 0.1 h = 4.0 h

Digital Data Capture

It shall be assumed that this project will be compiled with softcopy stereo equipment and methods. Therefore, diapositives will not be required. The film will be scanned for incorporation into the softcopy system during the aerotriangulation process. The map shown in Figure A.1 indicates that the project area will require planimetric and topographic feature compilation in portions of 40 stereomodels. The map also notes that of the 40 models an equivalent of approximately 29 full stereomodels will be collected. This can be determined by tabulating the total number of full models that would fill the project area. Review of the project area indicates that the planimetric data capture will be moderate to light. The town of Elsah, IL is the only area with any significant planimetric complilation. Road networks are limited to rural roads and highways. Water bodies will be limited to creeks and minor streams. No large water bodies are noted within the mapping area. It is assumed that 2 of the total 40 models will have moderate planimetric feature compilation. The remainder will be light. The topographic detail within the project is rolling to hilly. The hilly areas are in the southwestern portion (in the vicinity of the town of Elsah) and along the edges of the streams. It is assumed that approximately 20 models will have hilly topography and 20 models will have rolling topography.

Table A.1 itemizes the estimation of production hours required to collect planimetric data, while Table A.2 does the same for topographic data. Listed in Table A.3 is a summary of the production hours itemized above. Note that in addition to the total labor hours an appropriate overhead should be established and applied to the total cost of labor. Also, an appropriate profit should be established and applied to the total of labor and direct costs.

The process described above is to develop a rough estimate of the cost of a project for planning, negotiating, and budgetary purposes. Units and unit costs may vary and change based upon economics and equipment used by contractors. As equipment is improved, the time required for specific tasks may be reduced. However, the cost of the equipment and the person to operate the equipment may increase. By applying wage rates to labor hours and equipment rental to equipment hours the estimator can arrive at project budgetary costs. To support the process of estimating production hours of various photomapping phases, the estimator of a project is urged to seek current labor and equipment rates more applicable to a project or geographical area, perhaps using regional wage rates and actual negotiated contractor rates from recent or current contracts.

Table A.1 Summary of Productions Hours Required to Produce Planimetric Data

Planimetry				Approximate Planimetric Time (Hours/Model)			
				Final Map Scale			
Density Type	Models/Type	Hours/Type	Total Planimetry Hours	1 in. = 40 ft to 1 in. = 60 ft	1 in. = 100 ft to 1 in. = 150 ft	1 in. = 200 ft to 1 in. = 300 ft	1 in. = 400 ft to 1 in. = 1600 ft
Light							
1	38	2.5	95	3.0	2.5	2.5	2.5
2				4.0	3.5	3.5	3.5
Medium							
3	2	4	8	5.0	4.0	4.0	4.0
4				7.0	6.0	6.0	5.0
Heavy							
5				10.0	8.0	7.0	6.0
Total planimetry hours			103				
Edit time: generally 30% of total planimetric compilation hours			31				

Table A.2 Summary of Production Hours Required to Produce Topographic Data

Topography (Topo) Collection of Mass Points and Breaklines for Production of Contours				Approximate Topography Time (Hours/Model)			
				Final Map Contour Interval Scale			
Terrain Character (Slope)	Models/Type	Hours/Type	Total Topo Hours	1 ft	2 ft	4 ft	5–8 ft
Flat				2.0	2.5	2.5	2.0
Rolling	20	4	80	4.0	4.0	4.0	3.0
Hilly	20	6	120	6.0	6.0	5.0	4.0
Steep				8.0	8.0	6.0	5.0
Disturbed				10.0	10.0	8.0	7.0
Total topo hours			200				
Edit time: generally 30% of total topo collection time			60				

Table A.3 Summary of Direct Costs of Photogrammetric Mapping Project Production

Photogrammetric Mapping Project Production			
	Hours	**Unit Cost**	**Total Cost**
Production labor			
Aerial photography	10.1		
Aerotriangulation	40		
Model setup	0.44		
Planimetry	134		
Topography	260		
Orthophotography	N/A		
Total			
Direct Costs			
Film	44	Foot	
Prints	88	Each	
Diapositives	N/A	Each	
Hardcopy prints	N/A	Each	
CDs, disks, or tapes	2	Each	
Aircraft w/camera	2	Hours	
Stereoplotter	N/A	Hours	
Softcopy workstation	303	Hours	
Edit workstation	91	Hours	
Scanner	13.2	Hours	
Total direct cost			

SUGGESTED READING

Brown, C. M., *Boundary Control and Legal Principles,* 2nd ed., John Wiley & Sons, New York, 1969.

Ciciarelli, J. T., *A Practical Guide to Aerial Photography* (with an Introduction to Surveying), Van Nostrand Reinhold, New York, 1991.

Falkner, E., *Aerial Mapping Methods and Applications,* Lewis Publishers, Boca Raton, FL, 1995.

Hohl, P., Ed., *GIS Data Conversion: Strategies, Techniques, Management,* Onward Press, Santa Fe, NM, 1998.

Kavanagh, B. F., *Surveying with Construction Applications,* Prentice-Hall, Englewood Cliffs, NJ, 1989.

Macguire, D. J., Goodchild, M.F., and Rhind, D.W., *Geographical Information Systems: Principles and Applications,* Vols. 1 and 2, Longman Scientific & Technical, Essex, England, 1991(co-published in United States by John Wiley & Sons, New York).

Pratt, W. K., *Digital Image Processing,* John Wiley & Sons, New York, 1991.

Staff, Radar Basics, Canada Centre for Remote Sensing.

U.S. Army Corps of Engineers, Photogrammetric Mapping, EM 1110-1-1000, U.S. Army Corps of Engineers, 1993.

Wolf, P. R., *Elements of Photogrammetry,* McGraw-Hill, New York, 1983.

Index

C

D

E

F

G

H

I

K

L

M

N

O

P

Q

R

S

T

U

V

W